Micro Newton Thruster Development

Franz Georg Hey

Micro Newton Thruster Development

Direct Thrust Measurements
and Thruster Downscaling

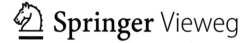

Franz Georg Hey
Friedrichshafen, Germany

Dissertation, Dresden University of Technology, 2017 with the title "Development and Test of a Micro-Newton Thruster Test Facility and Micro-Newton HEMPTs"

ISBN 978-3-658-24865-9   ISBN 978-3-658-21209-4   (eBook)
https://doi.org/10.1007/978-3-658-21209-4

Springer Vieweg
© Springer Fachmedien Wiesbaden GmbH 2018
Softcover re-print of the Hardcover 1st edition 2018

This Springer Vieweg imprint is published by Springer Nature
The registered company is Springer Fachmedien Wiesbaden GmbH
The registered company address is: Abraham-Lincoln-Str. 46, 65189 Wiesbaden, Germany

Acknowledgements

This work has been performed in the Laboratory for Enabling Technologies of Airbus in Friedrichshafen in close collaboration with the Dresden University of Technology and the University of Bremen / DLR Bremen.

Therefore, I want to thank my colleagues at Airbus, in particular Ulrich Johann, Dennis Weise, Noah Saks, Karlheinz Eckert, Ewan Fitzsimons, Hans Reiner Schulte, Andreas Keller, Harald Kögel, Oliver Mandel, Max Vaupel and Marjetka Kastner.

From the Dresden University of Technology, I want to thank Martin Tajmar.

I would like to acknowledge the support of Martin Gohlke, Thilo Schuldt, Tim Brandt and Claus Braxmaier in Bremen.

The Radio Frequency Ion Thruster tested has been provided by the ArianeGroup, thanks go to Christian Altmann and Hans Leiter, Marcel Berger.

My special acknowledgement goes to Günter Kornfeld, whose idea it was to use a cusp structured magnetic field as a thruster.

Lastly, I would like to thank Karolin and my family for their support.

Franz Georg Hey

Contents

List of Figures

List of Tables

List of Acronyms

ADC Analogue Digital Converter

AIAA American Institute of Aeronautics and Astronautics

AOCS Attitude and Orbit Control System

AOM Acousto Optic Modulator

ASD Amplitude Spectral Density

BS Beam Splitter

CAD Computer Aided Design

CoG Centre of Gravity

COTS Commercial Off The Shelf

CV COOLVAC

DAC Digital Analogue Converter

DAQ Data Acquisition Board

DC Direct Current

DDS Direct Digital Synthesiser

DFACS Drag Free Attitude Control System

DUT Device Under Test

DWS Differential Wavefront Sensing

EBB Elegant Bread Board

EM Engineering Model

EP Electric Propulsion

ESA European Space Agency

ESC Electro Static Comb

ESO European Southern Observatory

ESTEC European Space Research and Technology Centre

FEEP Field Emission Electric Propulsion

FEM Finite Element Method

FIFO First In First Out Memory

FPGA Field Programmable Gate Array

GIT Gridded Ion Thruster

GPS Global Positioning System

GW Gravitational Waves

HEMPT High Efficiency Multistage Plasma Thruster

HV High Voltage

LET Laboratory for Enabling Technology

LIF Laser Induced Fluorescence

LISA Laser Interferometer Space Antenna

LUT Look Up Table

NASA National Aeronautics and Space Administration

NGGM Next Generation Gravity Mission

NSSK North South Station Keeping

OLV Leybold Vacuum

PBS Polarising Beam Splitter

PCB Printed Circuit Board

PPU Power Processing Unit

PSD Power Spectral Density

PTTR Power To Thrust Ratio

QPD Quadrant Photodiode

RF Radio Frequency

RFG Radio Frequency Generator

RIT Radio Frequency Ion Thruster

RPA Retarding Potential Analyser

TED Thales Electron Device

TRL Technology Readiness Level

1 Introduction

Since the days Tsiolkovsky, Oberth and Goddard started dreaming of space exploration more than a century has passed. Today, the majority of the people are using
technologies which are based on space exploration such as satellite television or
satellite navigation and the space business plays a very important role in daily life.
But there are still a lot of challenging tasks to do and a lot more questions to answer.
One of the technologies to master the challenges is called Electric Propulsion (EP).
Already in 1906 the first basic idea of using electrical energy to generate a large
velocity of the propellant particle was born [1]. Since that time different technologies
for various applications have been developed and especially in the last years more
and more spacecraft with a so called EP system were used. Due to the development
of solar cell panels, highly efficient and highly miniaturised high voltage supplies
and the availability of sufficient computation power for plasma simulations the
maturity of EP systems has reached a high Technology Readiness Level (TRL) [2,
3]. Meanwhile missions like Dawn, GOCE and Hayabusa demonstrated the unique
advantages of an EP system for scientific space missions [4]. Moreover, electric
thrusters are used for North South Station Keeping (NSSK) and orbit raising of
telecommunication satellites.

1.1 Missions for Micro-Newton Thruster Applications

Almost at the time where the first ideas for EP were published (1911-1915), Albert
Einstein postulated the revolutionary theory of general relativity [5]. Most of the
predicted phenomena have been experimentally verified. One of the postulations
that was only recently verified is the existence of Gravitational Waves (GW) [6]. The
GW are variations in space-time. In accordance with Einstein's theorem that nothing
is faster than the speed of light, also gravity propagates with this velocity. Therefore,
a variation of the mass of a cosmic body would lead to this fluctuation inside space-
time. Possible sources are rotating neutron stars or black hole mergers. Typically,
the amplitudes of the GW are extremely small, hence the direct measurement of
GW is challenging.

With instruments that are placed on the Earth's surface, it is complicated to detect
these waves because of the significant seismic noise background in the environment.

© Springer Fachmedien Wiesbaden GmbH 2018
F. G. Hey, *Micro Newton Thruster Development*,
https://doi.org/10.1007/978-3-658-21209-4_1

Therefore, first plans for a space based gravitational wave detector were proposed in the 1980s [7]. Since the year 2000 Airbus in Friedrichshafen is working on the mission concepts for such a scientific space mission, the so called LISA. The recent concepts aim to detect GW in a frequency band between 30 µHz and 1 Hz [8, 7].

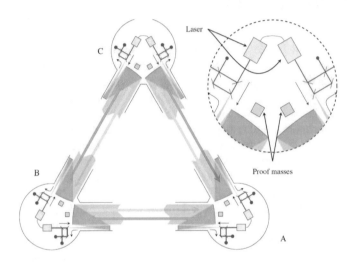

Figure 1.1: Sketch of the LISA constellation, with six identical laser ranging terminals, two per spacecraft. Each terminal houses its own test mass. The measured translation of the test mass versus the satellite structure is the input signal for the Attitude and Orbit Control System (AOCS) control loop [7].

The classic LISA concept consists of three identical satellites which forms a three armed interferometer in space. The constellation allows measuring the variation of the space between the satellites in the pico-Meter regime. The constellation is presented in figure 1.1. A single interferometer arm consists of two proof masses which are free flying inside the satellites. Each proof mass has a separate laser interferometer system which measures the position of the proof mass versus the satellite structure. A bidirectional laser interferometer link between the satellites measures the position of the satellite structures in relation to each other. Thus, the relative distance variation of the test masses can be determined. The laser link between the satellites is established via telescopes. The moveable telescopes are used to transmit and receive the laser beams. The satellites form an equilateral triangle in a heliocentric orbit. The formation follows the Earth's orbit within 20 °, which corresponds to approximately 50 million kilometres. Figure 1.2 illustrates the LISA science orbit. The orbit enables an optimal data transmission to Earth and power generation as well as low gravitational effects caused by the Earth and other space bodies. Moreover, the ΔV required to reach the final orbit is minimised.

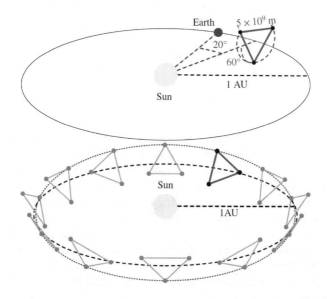

Figure 1.2: Sketch of the LISA orbit, the satellites follow the Earth orbit within 20°, or in approximately 50 million kilometres. The inclination with respect to the ecliptic is 60° [7].

To reach the required measurement resolution, the free flying proof masses have to be shielded against external disturbance such as the radiation pressure, or the solar wind.. The satellite structure itself acts as the shield. Therefore an active steering of the structure is required. The special AOCS, or steering system, is called Drag Free Attitude Control System (DFACS). The permanent cancellation of all non-gravitational effects must be performed with extremely precise actuators that should be able to compensate the non-symmetric disturbance with high precision and with a wide dynamic range.

The actuators will be controlled by a laser interferometer system that measures the proof mass distance with respect to the satellite structure [8, 7, 9]. According to the requirements, only a micro-Newton thruster system can be used, especially due to the asymmetry of the disturbances mentioned. Mainly, the thrust generated by the thrusters is used to counteract the solar radiation pressure that amounts to about $10\,\mu N$ [7, 9]. With respect to the requirements of the laser instruments the

Table 1.1: Summary of the LISA micro-Newton thruster requirements [7, 9].

Parameter	Requirement
Maximum thrust	$100\,\mu N$
Bandwidth	$10\,Hz$ to $2 \cdot 10^{-3}\,Hz$
Thrust noise	$0.1\,\mu N/\sqrt{Hz} \cdot \sqrt{1 + \left(\frac{10\,mHz}{f}\right)^4}$
Lifetime	$> 5\,yr$
Total impulse	$7800\,Ns$ per year

thrust noise has to be below $0.1\,\mu N$. The requirements are summarised in table 1.1. Figure 1.3 illustrates the required thrust noise level (dash-dot-dotted curve) in comparison with other future scientific space mission requirements which will also be discussed. In the context of LISA it has become common to refer to the spectral density given in $1/\sqrt{Hz}$ as Power Spectral Density (PSD). Formally, this is a linear spectral density (also Amplitude Spectral Density (ASD), i. e. $PSD^{1/2}$) [8]. Hence, in the following thesis PSD is used in the described manner. PSDs are describing the distribution of noise components in the frequency space of a measured signal. They can be computed according to Fourier analysis. The PSD which are presented in the following are computed according to the summarised information from G. Heinzel et. al. [10].

In parallel to LISA several other scientific space missions are in preparation. In particular, future space telescope concepts as well as formation flying Earth observation missions have a need for highly precise AOCS. Mission examples are the Darwin space telescope, the Next Generation Gravity Mission (NGGM), and the Euclid space telescope.

The Darwin mission is a planned infrared space telescope to observe Earth-like exoplanets. The concept is for a nulling interferometer in space which consists of up to six $1.5\,m$ class telescopes arranged in a hexagonal configuration. Figure 1.4 illustrates the Darwin satellite formation. Each telescope will be carried by a single satellite. In the middle of the configuration a mother satellite will be placed. Highly precise laser links between the satellites enable the nulling and the imaging of the observed target. Just like LISA, the solar radiation pressure has to be compensated. The position of the satellites needs to be stabilised in the range of $5\,nm$. Therefore a micro-Newton propulsion system is required. The requirements for the micro propulsion system of Darwin are almost identical to the requirements of LISA. Therefore the dash-dot-dotted curve that is plotted in figure 1.3 is also representing the thrust noise requirement of Darwin.

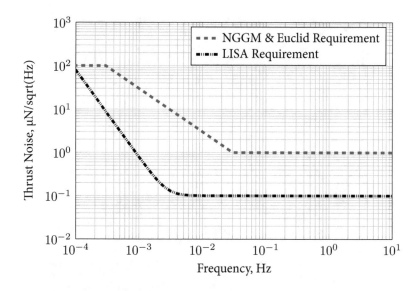

Figure 1.3: The PSD summarises the thrust noise requirements of NGGM, Euclid and LISA. The required thrust noise level for NGGM and Euclid is $1\,\mu N\,/\sqrt{Hz}$, whereas LISA requires $0.1\,\mu N\,/\sqrt{Hz}$ [11, 12, 7, 9].

As mentioned, NGGM is a planned future space mission which needs micro-Newton propulsion. The actual concept consists of two satellite pairs. The first pair will be placed in a circular, low and polar orbit around Earth. Whereas, the second pair will be operated at the same altitude on an inclined orbit. Highly precise accelerometers and an inter satellite laser link will lead to observations of the gravity field of the Earth over a long time span, with a high spatial and temporal resolution. This allows detailed investigations of geophysical phenomena of the Earth. The altitude of the orbit will be in the range of 300 km - 400 km. Thus, the drag from the residual atmosphere is around 2 mN [11]. Moreover, all other non gravitational acceleration have to be compensated which will require thrust levels in the order of micro-Newtons [11]. The thrust noise requirement of the mission is plotted in figure as 1.3 as the dashed curve.

A representative example for a single satellite that requires highly precise AOCS is the Euclid space telescope. The instruments of the satellite will create a map of the universe in the infrared spectrum. This will allow detailed studies of the dark matter and dark energy distribution within the universe [12]. To achieve the targeted mission goals, the propulsion system has to provide two different operation modes. A highly stable observation mode and a stepping mode [14, 12]. Especially in the

Figure 1.4: Artist's impression of the Darwin configuration, the six telescopes are surround-
ing the mother satellite and forming a Nulling-Interferometer. A data relay
satellite is shown as well [13].

observation mode the thrust noise of the propulsion system is important. The thrust
noise requirement of Euclid is similar to the thrust noise requirement of NGGM,
hence the requirement is also presented in figure 1.3 as the dashed curve. [14, 11].

In the last years small satellites and especially small satellite constellations have
received a lot of publicity. Famous examples are the SpaceX satellite constellation
that would consist of up to 4000 satellites [15], or the OneWeb satellite constellation
which will consist of up to 720 satellites [15, 16]. The main goal of the companies
involved is to provide global internet access, particularly for underdeveloped areas.
Other companies like Virgin, Facebook and Google are also interested in this type
of business.

The spacecraft that will potentially form the constellation will probably have a
mass in the range between 20 kg to 150 kg. Due to the targeted application, the
antennas used must always point to Earth, therefore the satellites have to be 3-axis
stabilised. Furthermore, the satellites will operate in a low Earth orbit and hence a
compensation of the atmospheric drag is required. Moreover, the propulsion system
has be used for the orbit raising of the spacecraft as well, in order to minimise the

launch costs of the constellation. Furthermore, a graveyarding of the satellites have to be performed. Because of the small satellite mass, the propulsion system used has to be extremely efficient, but also simple and cheap.

In general, different types of thrusters and other actuators could be used to fulfil the requirements of the applications mentioned. In the past, cold gas thrusters were used especially for micro-Newton propulsion applications. But, for the given mission examples, more efficient propulsion systems like electric thrusters would have a lot of benefits compared to the traditional technologies such as cold gas systems or chemical thrusters. In comparison, the EP technology efficiency is usually 10 to 100 times higher, dependent on the specific EP type. Typically, an expansion of pressurized gas into the vacuum, which is termed a cold gas thruster, has a specific impulse between 30 s to 90 s. Chemical combustion like in a rocket can achieve specific impulses between 200 s and 468 s. Whereas, EP systems are able to achieve specific impulses of more than 3000 s [17, 2]. This leads to significant savings of the overall required propellant mass, if an EP system is used instead of a traditional system. Hence, the satellites with EP can be operated much longer with the same amount of propellant. This is especially important for the future scientific space missions mentioned. Other benefits for those missions are the wide throttle range and the low thrust noise of electric thrusters. The efficient use of propellant can also lead to significant overall spacecraft mass savings [18] that allows the designing of small and cost efficient spacecraft or an increased payload mass.

1.2 Micro Propulsion Thrusters Overview

There are several options for micro-Newton electric thrusters. However, up to now only cold gas systems have been flown into space. But, due to the advantages of an EP system, different electric thrusters are in development. Thruster technology examples that could be used for highly precise AOCS are the RIT, Field Emission Electric Propulsion (FEEP), colloid and electrospray thrusters, or the µHEMPT. In the following chapters the focus will on RIT, FEEP, and µHEMPT. A simplified comparison between the technologies is given in table 1.2. Dependent on the specific concept the thrusters have different advantages and disadvantages, an overview of the thruster physics will be given in section 2.3. In general RIT systems have a high complexity and medium dynamic range, but low thrust noise and a long life time. Usually a FEEP system has a short lifetime and a low dynamic range, but a medium system complexity and low thrust noise. HEMPTs have typically a very long life time, a low system complexity and a high dynamic range, but the thrust noise of the thruster is not as good as the other thruster types due to the simple design.

Table 1.2: Qualitative tradeoff between the FEEP, RIT and HEMPT technology for a comparison of advantages and disadvantages.

Feature	FEEP	RIT	HEMPT
System complexity	Medium	High	Low
Dynamic range	Low	Medium	High
Thrust noise	Low	Low	Medium
Lifetime	Short - Middle	Long	Very long

However, all mentioned EP thruster types provide a sufficiently high dynamic range, excellent thrust noise properties and high efficiency compared to a cold gas system. The advantages of a cold gas system are the low system complexity compared to an electric thruster and the fact that cold gas systems have a lot of flight heritage. Cold gas thrusters are already qualified and used in the micro-Newton range in space on-board the Gaia space telescope. Up to now, no electric micro thruster has reached a TRL larger than five.

FEEP thrusters can be categorised by the propellant used. Typical FEEP propellants are Indium, Gallium, or Caesium. The development of FEEP thrusters using Caesium as propellant started more than 30 years ago [19]. Since 1995, Indium FEEPs are developed in Austria for highly precise AOCS [20]. Dresden University of Technology started the development of highly miniaturised Gallium FEEP thrusters in 2012 [21]. Typically, the thrust of a single FEEP thruster is in the order of tens of micro-Newtons, therefore it is especially interesting for precise AOCS applications.

The RIT was invented at the University of Gießen, Germany by Horst Löb in the 1960s. Since that time, the RIT development has been continuously running. The first in-flight experience of a RIT system was made on Artemis in 2001 [22]. Currently, different types of thrusters for a wide range of applications are in development, in the thrust range between 50 μN to 165 mN. A dedicated thruster based on the RIT technology that is specialised for highly precise AOCS is the RITμX. The thruster is in development since the early 2000s [23].

The HEMPT, sometimes called Cusped Field Thruster was invented by Günter Kornfeld at Thales Electron Device (TED) in 1998 [24]. Günter Kornfeld derived the basic features of the thruster from travelling wave tubes and their electron collectors, where the electron beam that is used to amplify a microwave signal is confined in a magnetic field, which has a cusp field configuration. The idea was to create a thruster where the plasma has almost no contact with the walls and therefore no life time restrictions caused by wall erosion such as in Hall-Effect Thrusters. TED is developing the thruster as a satellite main thruster for NSSK [25, 26, 27, 28, 29]. To develop a HEMPT that can be used in the micro-Newton regime, a parameter study was initiated and has been performed by Airbus since 2009 [30, 31].

The results of the downscaling parameter study performed were published in the PhD thesis of Andreas Keller [32]. The study compares different geometrical HEMPT configurations to identify the most important parameters. The measurements were performed at the University of Gießen. Within the study the first micro-Newton HEMPT was built and tested. The electric parameters were measured and analysed and a lot of important information for the further thruster development has been obtained. During A. Keller's study different techniques to characterise the thruster were used, such as Faraday cup measurements, RPA measurements and xenon excitation spectroscopy. However, no direct thrust measurement was performed because no sufficient micro-Newton thrust balance was available [32]. However, only a direct thrust measurement enables the assessment of all important thruster parameters and the overall thruster efficiency.

The μHEMPT developed was able to operate in a thrust range between $50\,\mu N$ and $900\,\mu N$. At lower thrust levels the specific impulse was in the range of $200\,s$. In summary, the study was able to wrap up the most important downscaling parameters but the developed thrusters were not able to fulfil the requirements for the targeted future scientific space mission as presented in section 1.1 and figure 1.3.

1.3 Micro Newton Test Facilities in Europe

Due to the complex physics of EP and the required qualification process for space applications, potential thruster candidates have to perform a wide range of tests before they are used in space. At the beginning of the thruster development as much information as possible has to be gained about the thruster via different kinds of tests. For instance direct thrust measurements, monitoring of the electric and physical thruster parameters, thruster plume diagnostics, etc. After the successful development of a thruster, it has to be characterised and qualified which means the detailed analysis of the thruster characteristic, documentation of all important thruster system parameters, life time testing, etc. This can be performed with the methods that are used on the thruster development. After a successful qualification of the thruster type, it can be used on a spacecraft. However, usually before the flight hardware is integrated, functional tests have to be performed to avoid potential malfunctions in space [33]. Therefore, every flight unit has to be tested on ground. For all these activities specific test facilities have to be available.

Currently, different kind of facilities are available in Europe. Facilities that are specifically dedicated for micro-Newton propulsion are located at the University of Gießen in Germany, at ONERA in Paris in France, at Dresden University of

Technology in Germany, at Fotec in Austria and at European Space Research and Technology Centre (ESTEC) in the Netherlands [34, 35, 36, 37, 38]. However, the specific capabilities of the most facilities are continuously changing, dependent on the test campaigns that are performed. Therefore, the following will not provide a full overview.

The University of Gießen is able to provide vacuum facilities where different plasma diagnostic setups are integrated, such as Faraday cup arrays, a RPA, or mass spectroscopy [34, 39, 32, 40]. Several vacuum tanks with high pumping capacities are installed and low background pressures can be achieved. Hence, reliable measurements with micro- and milli-Newton thrusters can be performed.

The French aerospace laboratory ONERA has different kinds of vacuum tanks as well. A highly precise thrust balance is available especially for micro-Newton propulsion [41, 35]. Moreover, a Laser Induced Fluorescence (LIF) setup is part of the Laboratory [42]. The ONERA micro-Newton thrust balance was used to characterise the Gaia micro-Newton cold gas thrusters [43, 44]. The balance is able to fulfil the Euclid and the NGGM requirements, however it does not fully comply with the LISA requirements (see figure 1.3).

Dresden University of Technology developed a highly precise thrust balance. The balance is placed in a vacuum tank [36]. The balance has been used to test laboratory models of different FEEP thrusters and other novel technologies.

Fotec in Austria has developed different kinds of thrust balances, especially for the testing of FEEP thrusters [38]. The actual balance has a measurement range from a few micro-Newtons to several milli-Newtons and can bear a propulsion system with a total weight of up to 13 kg. The thrust balance was also tested and verified at the ESA propulsion laboratory at Estec [45].

At Estec in the Netherlands various vacuum tanks and diagnostics are available [37]. For micro-Newton propulsion testing a thrust balance is part of the laboratory [46, 47]. The balance has a resolution of $0.2\,\mu N$. It was extensively tested with a cold gas thruster. However, no reliable test with a micro-Newton EP system has been performed.

In parallel to the mentioned parameter study of Andreas Keller, a small test facility was also built at Airbus in Friedrichshafen. Until 2012, the EP facility at Airbus in Friedrichshafen consisted of a 300 l vacuum tank, a turbo molecular pump, power supplies, and a propellant feeding system. The pump had a pumping speed of 700 l/s. The facility has been used to perform preliminary micro-Newton thruster tests and electron spectroscopy. But, due to limited capabilities of the facility, the thruster characterisation in the scope of the HEMPT downscaling feasibility study was performed at the University of Gießen [32]. In particular, the low pumping speed rendered the facility insufficient for reliable thruster characterisation because of the high neutral gas background pressure level. Except for the electron spectroscopy setup, no other instruments were part of the facility. However, the first development steps to develop an in-house micro-Newton thrust stand were undertaken [48].

It is clear that different test facilities exist in Europe and that a wide range of instruments for thruster characterisation are available. Hence, the European EP community is able to fulfil detailed characterisation, verification and qualification of electric thrusters in different thrust regimes. But, it is also clearly evident that especially for micro-Newton propulsion the test stands that are distributed all over Europe do not fully comply with the requirements of future scientific space missions such as LISA. Especially for the thrust noise characterisation of possible LISA thruster candidates, no suitable test facility was available until now.

1.4 Airbus Development Approach

Due to the experience that was gained in the parameter study of the µHEMPT, it became clear that a new micro-Newton thruster test facility was necessary to continue the independent micro-Newton thruster development for LISA-like missions at Airbus in Friedrichshafen. The existing thrust balances were not able to fulfil the requirements which were needed to characterise micro-Newton thruster candidates for LISA. In order to overcome this limitation a micro-Newton thrust balance should be developed as part of the facility. To maintain the continuity of the measured data from the parameter study performed, a basic set of plasma diagnostics, that is comparable to the instruments at the University of Gießen, should be part of the facility.

The approach, defining the goals of the work presented here, is divided into two main tasks. The first task was the development of a micro-Newton thruster test facility that fully complies within the requirements of the Euclid, the NGGM and the LISA missions. Therefore, the facility has to offer the possibility for highly precise direct thrust measurements to characterise the thrust noise of possible LISA AOCS thruster candidates as well as absolute thrust measurements of the investigated thrusters. Additionally, the facility has to offer the ability to measure the electric parameters of the propulsion system under test including basic plume diagnostics. Of course, the measurements must be performed under reliable measurement conditions.

The second task was the development of a next generation µHEMPT laboratory model which is able to demonstrate the unique advantages of the HEMPT technology as well as the operation space that is required for future scientific space missions such as LISA.

Specifically, the Airbus approach to uses the heritage in highly precise metrology [49, 50, 51, 52, 53, 54] and as a major scientific satellite manufacturer to develop a

Table 1.3: Summary of the goals of the micro-Newton thruster facility development. The goals are derived from the needs of the targeted missions and the previously gained experience.

Components	Parameter	Development target
Vacuum facility requirements	Vacuum chamber volume	$> 1000\,l$
	Background pressure with active thruster (0.5 sccm gas ballast)	$\approx 1 \cdot 10^{-6}\,mbar$
Thrust balance	Thrust range	$0.1\,\mu N$ to $2500\,\mu N$
	Measurement bandwidth	$10\,Hz$ to $2 \cdot 10^{-3}\,Hz$
	Thrust noise sensitivity	$0.1\,\mu N/\sqrt{Hz}$
	Maximum device under investigation mass	$6\,kg$
Other diagnostic	Plume characterisation options	Beam current, divergence, and energy measurements

highly stable and highly precise thrust stand and to use this instrument to accelerate the µHEMPT development in close cooperation with university research.

Table 1.3 is provides an overview of the goals of micro-Newton thruster test facility development. The goals are defined based on the needs of the targeted missions (LISA, NGGM, etc.) and the experience achieved by A. Keller's parameter survey. The minimum volume of the vacuum chamber is derived from the experience which was gained during the operation in smaller vacuum chambers. The background pressure follows the recommendation of the American Institute of Aeronautics and Astronautics (AIAA), the National Aeronautics and Space Administration (NASA) and the European Space Agency (ESA) [55, 56]. The requirements of the thrust balance are derived from the presented scientific space missions goals.

The goals of the thruster development can also be derived from the presented future space missions. Additionally, the specific HEMPT characteristics and the result that has been obtained by A. Keller are taken into account. The main thruster goals are:

- Thrust range from $50\,\mu N$ to $250\,\mu N$

- Demonstrated thrust noise of $0.1\,\mu N\,/\sqrt{Hz}$

- Specific impulse at all operation point $\geq 1000\,s$

With respect to the presented information and the previously performed work of A. Keller, this paper will present and summarise the Airbus approach to overcoming the current limitations of micro-Newton electric thruster characterisation and testing to the LISA requirements as well as the further µHEMPT development on its way to TRL 3 [3].

2 Electric Propulsion Fundamentals

Electric thrusters use the same principles as chemical rockets to propel spacecraft, however the ejected mass consists basically of charged particles. Furthermore, the acceleration of the propellant particles is not powered by the energy stored in the propellant, but rather by an external power source. This changes the way the parameters of the thruster are calculated. Due to the decoupling of the energy source from the propellant, EP can provide much higher exhaust velocities than chemical rockets. Those generally have exhaust velocities of about $3\,\mathrm{km/s}$ to $4\,\mathrm{km/s}$ while the exhaust velocities of EP thrusters can reach up to $100\,\mathrm{km/s}$ [17].

Since the first ideas of how to go to space were published, a lot of different kinds of thrusters were developed, they are all based on the principle of linear momentum conservation. As long as the spacecraft and the accelerated matter are not as fast as the speed of light, Newton's classical equation of linear momentum can be used. The total impulse that a thruster can generate (p_t) is equal to the total propellant mass (m_t) multiplied by the relative velocity (v_p) of the propellant particles [17, 2].

In this publication only electric thrusters that use an applied electric field to accelerate the ionised propellant particles are discussed. Hence, the words electric thruster, EP, ion thruster, or ion source are used equally in this meaning.

The thrust produced is proportional to the velocity of the ionised particle ($v_i = \sqrt{2 \cdot q \cdot U_b / m_{ion}}$), as the square root of two times the particle charge (q) multiplied by the beam potential (U_b) divided by the mass of a propellant ion m_{ion} [17, 2].

The correlation between the total propellant mass and the velocity allows the derivation of the general goals of the thruster development. One major goal is the reduction of the overall system mass via the minimisation of required propellant mass for a given total impulse, this can be easily achieved with EP. As mentioned, EP requires an external power source. Therefore, also the power consumption of the thruster has to be considered. Dependent on the spacecraft mass the available power varies from only a few watts, for single unit CubeSats [21], to several kilowatts, for telecommunication satellites [17, 2]. According to the available power, electric thrusters are typically able to provide maximum thrust levels of hundreds of milli-Newtons. Therefore, EP thrusters require very long firing durations, compared to chemical thrusters, to create the same total impulse.

© Springer Fachmedien Wiesbaden GmbH 2018
F. G. Hey, *Micro Newton Thruster Development*,
https://doi.org/10.1007/978-3-658-21209-4_2

It becomes clear that to develop, design and operate an EP system as opposed to a chemical system, different key figures, a different system design and different operation strategies are necessary [17, 2].

2.1 Basic Equations

To compare different thruster technologies, standardised key figures are defined. For EP the most important figures are the specific impulse (I_{sp}) and the Power To Thrust Ratio (PTTR). The specific impulse describes the propellant efficiency of a thruster and is defined as the thrust (F_t) divided by the mass flow of propellant ($\dot{m}$) [2]. Therefore, the I_{sp} would have the dimension of a velocity, however the specific impulse is normalised by the acceleration at the Earth's surface (g_0) to improve the comparability with different unit systems and is therefore measured in seconds. The I_{sp} can be written as

$$I_{sp} = \frac{F_t}{g_0 \cdot \dot{m}} = \frac{\gamma \cdot \eta_m}{g_0} \cdot \sqrt{\frac{2 \cdot q \cdot U_b}{m_{ion}}} \quad , \tag{2.1}$$

where γ is the beam divergence correction factor for the thrust directional velocity, η_m is the mass efficiency correction factor or ionisation efficiency, U_b is the beam acceleration voltage, q is the charge of the ions and m_{ion} is the mass of a single propellant particle [2].

The equation underlines that a reduction of the mass flow at a constant thrust level induces a higher specific impulse. One way to decrease the mass flow at a constant level of thrust is to increase the propellant speed. Dependent on the energy source of the propulsion system, different propellant velocities (v) and therefore different specific impulses can be achieved. The electric thrusters discussed operate specific impulses between $1000\,\text{s}$ and $8000\,\text{s}$ [17].

By estimating the specific impulse with the electrical parameters of the thruster as described by the right part of equation 2.1 losses have to be taken into account. The beam voltage can be calculated through the voltage efficiency, sometimes called acceleration efficiency and the anode voltage ($U_b = \eta_v \cdot U_a$). Typically, in electric thrusters not all propellant particles are ionised and hence it is necessary to determine the mass utilization efficiency ($\eta_m = I_b \cdot m_{ion}/q \cdot \dot{m}$), which illustrates how many neutral particles are ionised inside the thruster.

To take into account that usually the ions leaving the thruster also have a velocity perpendicular to the thrust direction, the divergence efficiency (γ) has to be determined. It can be calculated as the cosine of the average divergence half-angle, called θ, of the beam.

The thrust and the specific impulse can be determined exclusively with electrical parameters, i.e. with the anode voltage and the anode current (I_a). However, a basic set of efficiency figures, such as the discharge efficiency (η_d), the electrical efficiency (η_e), the acceleration efficiency (η_v) and the already introduced divergence efficiency (γ), is required to perform this determination.

The anode current can typically be determined with the power supplier used. This current can be transferred into the current of the beam ($I_b = I_a \cdot \eta_d$) with the discharge efficiency η_d.

By assuming that the overall thruster power is only the anode current multiplied by the anode voltage ($P_T = V_a \cdot I_a$), the electrical efficiency is the product of the acceleration efficiency and the discharge efficiency ($\eta_e = \eta_d \cdot \eta_v$).

The key figures introduced allow a calculation of the total thruster efficiency which is given by

$$\eta_T = \gamma^2 \cdot \eta_e \cdot \eta_m \quad . \tag{2.2}$$

If the total efficiency of the thruster is know, the thrust can directly be calculated with the input parameters of the thruster. It is also possible to derive the total efficiency of the thruster if these parameters and the thrust are known, for example via a direct thrust measurement. Therefore, the total efficiency can be written as

$$\eta_T = \frac{F_t^2}{2 \cdot \dot{m} \cdot P_t} \quad . \tag{2.3}$$

According to equation 2.1, the specific impulse depends only on the mass of a single particle and the acceleration voltage. In first instance, the connection between the specific impulse and the acceleration voltage seems to be attractive to optimise the efficiency of an EP thruster. It can be seen that the use of lighter ions will increase the I_{sp}. As mentioned above, electric propulsion always requires an external energy source and therefore the power has to be taken into account to fully understand the relationships between the specific impulse, beam voltage, mass flow and particle mass.

The specific kinetic power of the particle jet leaving the thruster is called jet power, and is defined as

$$P_{jet} = \frac{1}{2}\dot{m} \cdot v^2 \quad . \tag{2.4}$$

With the assumption that only ionised particles leave the thruster the jet power is equal to the power of the beam ($P_{jet} = P_b$), the beam power becomes

$$P_b = \frac{F_t^2}{2 \cdot \dot{m}} = \frac{g_0^2 \cdot \dot{m} \cdot I_{sp}^2}{2} \quad . \tag{2.5}$$

This expression shows that an increased thrust at a constant mass flow will increase the beam power and it underlines that the required beam power would increase with the square of the specific impulse.

Due to the external energy source, required for EP, for a detailed comparison of different types of EP thrusters an additional key figure is required, in parallel to the specific impulse. The usage of the relation between overall required power, so called total power (P_t) and the thrust produced is common practice. It allows a simple, fast and uncomplicated comparison of different thruster types. The so called PTTR is given as

$$\text{PTTR} = \frac{P_t}{F_t} \qquad . \tag{2.6}$$

$$\tag{2.7}$$

In practice, low $PTTRs$ are required due to the limited amount of power on the spacecraft. Typically, HEMPTs have a PTTR between $20\,\text{W/mN}$ and $30\,\text{W/mN}$, whereas RITs have a PTTR of around $33\,\text{W/mN}$ [2, 57] and FEEPs are in the range of $70\,\text{W/mN}$.

In general, higher specific impulses generate a higher PTTR, i.e. the required power will increase for a certain amount of thrust to achieve a higher specific impulse. As mentioned above, lighter propellant particles would also lead to a higher specific impulse, combined with a higher PTTR. Therefore, heavier propellants, such as xenon, are preferred to produce as much thrust as possible with the available power on the spacecraft.

The most precise way to determine the PTTR is to directly measure the thrust produced and in parallel the total input power, i.e. all losses were directly taken into account. Within the parallel measurement of the input mass flow also the I_{sp} can be derived. This method is called direct thrust measurement.

2.2 Basic Plasma Physics

Using electrical energy to achieve high specific impulses leads to the concept of acceleration of charged particles in an electric field. These charged particles are generated from a propellant, in most cases a gas. The neutral gas particles are separated into ions and electrons to create what is called plasma. The charged particles move inside the plasma due to the electric and magnetic fields that are created from the particles themselves, or which are applied from externally. Usually the number of electrons and ions are almost equal, i.e. from an outside perspective the plasma is quasi neutral. The ions and electrons can be characterised by their temperatures (T_i and T_e), which are usually not the same. The force between two charged particles can be calculated with Coulomb's law.

Moreover, the motion and the behaviour of charged particles can be described by Maxwell equations and the Lorentz force equation. Commonly, Maxwell equations are used to describe the electric and magnetic field inside an electric thruster whereas the Lorentz force equation is used to propagate the particle through the electric and magnetic field. The Lorentz force is

$$\vec{F_l} = q \cdot (\vec{E} + \vec{v} \times \vec{B}) \quad . \tag{2.8}$$

From this formula the motion of a particle in an E-field respectively a B-field can be derived. For example, a particle would follow the electric field lines directly, the direction of the particle motion depends on the specific charge of the particle [2]. The Lorentz force caused by the magnetic field acts perpendicularly to the velocity of the particle and the magnetic field lines. Thus, when a particle that has a specific velocity enters a uniform magnetic field it will start a cycloidal motion perpendicular to the magnetic field direction and the velocity direction. The corresponding centripetal force is

$$\vec{F_c} = q \cdot (\vec{v_\perp} \times \vec{B}) = \frac{m_p \cdot v_\perp^2}{r} \quad , \tag{2.9}$$

where m_p is the particle mass and r is the radius of the circle,

$$r = r_l = \frac{m \cdot v}{q \cdot B} \quad , \tag{2.10}$$

which is called the Larmor radius. By taking the particle velocity into account the Larmor radius can be written as

$$r_l = \frac{1}{B} \sqrt{\frac{2m \cdot U}{q}} \quad . \tag{2.11}$$

The equations illustrate the basic relationships between the magnetic field strength, the particle mass, the particle charge and the particle velocity. For example, a stronger magnetic field leads to a smaller Larmor radius, a higher velocity causes a larger Larmor radius. As mentioned, above the plasma consists of ions and electrons. A single ionised ion has the same charge as an electron, but the mass difference between an electron ($m = 5.457 \cdot 10^{-4}$ u) and an ion (e.g. a xenon ion, $m = 131$ u) is almost six orders of magnitude, i.e. the Larmor radius of a xenon ion, at constant charge, velocity and B-field, is 1000 times larger than the Larmor radius of an electron [2].

The significant mass difference of the particles in the plasma leads to a general different behaviour of the particles. According to equation 2.8, the electrons are extremely agile, i.e. small force and therefore small E-Fields and B-Fields can cause large changes of the motion of single electrons. In contrast to the electrons, a strong E-Field and B-Field are required to change the motion of an ion.

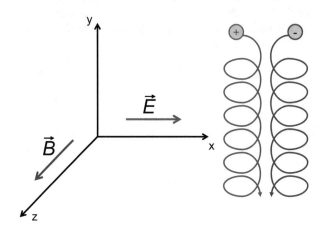

Figure 2.1: Illustration of the typical ExB drift of charged particles inside an electrical field and a magnetic field. The gyration direction of the particle depends their charge [58].

The motion of a charged particle in a volume with a finite E- and B-field is called ExB drift, in reference to equation 2.8, with a negligible δt it can be described as

$$E = -v \cdot B \qquad . \tag{2.12}$$

By taking the cross product of B on both sides the equation can be transformed into

$$v_{ExB} = \frac{E \times B}{B^2} \qquad . \tag{2.13}$$

Figure 2.1 presents the motion of a singly charged ion and an electron in a uniform B- and E-field. The particles spin around the magnetic field lines, but in opposite directions because of their opposite charge. The rotation velocity and the radius depends on the field strengths.

In electric thrusters, especially in a HEMPT, the magnetic field configuration is not uniform. It is usually changing in magnitude along the magnetic field lines. If the magnitude of the B-field increases, according to equation 2.11, the Larmor radii of particles that are bonded to a magnetic field line would decrease. Simultaneously, the distance between the field lines decrease, thus the field lines deviates not parallel.

Therefore, the Lorentz force no longer acts perpendicularly to the drift direction and contains a component that acts against the drift direction of the particle. The particles accelerate versus the drift direction, i.e. it is possible that the particles slow down until they have zero velocity in the drift direction, or they could even start drift into the opposite direction. This means that a rotational kinetic energy component of the particle can be transferred into an axial velocity component and vice versa. This field configuration is called a magnetic mirror.

Figure 2.2 illustrates the magnetic field configuration described. The sketch presents the B-Field lines in z-direction in a so called magnetic bottle and a particle (in red) which gyrates inside the magnetic field. Along the z-axis the strength of the magnetic field increases, thus $\Delta B \neq 0$ and the magnetic field lines become more dense. Hence, the particle shown is accelerated versus its drift direction and is reflected in the area with the higher B-Field density due to the gradient of the B-Field.

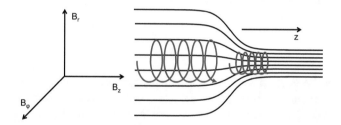

Figure 2.2: Visualisation of a magnetic bottle. The particle (in red) which gyrates in parallel to B-Field underlies a force that acts perpendicular to the drift direction due to ∇B and thus the particle can be reflected by this magnetic field topology [58].

During the motion of the particle through the magnetic field, the energy is conserved because the B-field vector remains perpendicular ($|v_0| = |v_m|$). Therefore, the magnetic mirror can be described by the relation of the B-field magnitude at the start point of the particle (B_0) and the B-field magnitude inside the magnetic mirror (B_m) and is given as

$$\frac{B_0}{B_m} = \frac{v_{\perp,0}^2}{v_0^2} = sin^2(\Theta_m) = \frac{1}{R_m} \quad , \tag{2.14}$$

where $v_{\perp,0}$ is the perpendicular velocity at the starting points, v_0 is the absolute value of the velocity vector, Θ_m is the pitch angle inside the magnetic mirror of a charged particle between the particle's velocity vector and the local magnetic field and R_m is the mirror ratio. The pitch angle is a measure that describes if a particle is trapped in the magnetic mirror or not. From equation 2.14 the reflection constraint can be formulated:

$$v_{\parallel} < v_{\perp} \sqrt{R_m - 1} \qquad . \tag{2.15}$$

The equation can be used to provide information about the confinement of energetic electrons in electric thrusters. Whenever charged particles are distributed over a region of space the charge between the particles plays an important role. The interaction between the particles are described by Coulomb's Law.

As mentioned, an ideal plasma is neutral, which means that the number of positive and negative particles/charges is equal and ideally the coulomb forces inside the plasma would be negligible. A tiny displacement of electrons causes a Coulomb force as a restoring force, but due to the mass difference between the electrons and the ions, mainly the electrons are pulled back. Thus, the plasma oscillates. The frequency of the electron oscillation is defined by

$$\omega_p = \sqrt{\frac{n_e \cdot e^2}{\epsilon_0 \cdot m_e}} \qquad , \tag{2.16}$$

where n_e is the electron density, e is the elementary charge, ϵ_0 is the permittivity of free space and m_e is the electron mass. It is called the electron plasma frequency.

The charge differences and the fields inside the plasma can also cause areas where one species dominates. In this case a space charge would be formed which also means that the plasma potential in an area with a dominant species is different from the other areas.

From the Poisson equation, it can be explained why an EP device always requires a neutralisation of the ion beam. To be specific, if only ions were be extracted from the spacecraft, a negative space charge would be formed around the thruster and the spacecraft and no further ions could be extracted.

Based on the presented fundamental equations, different thruster types will be explained in the following section. In chapter 4 the information provided will be used to explain the thruster development which is also part of this thesis.

2.3 Electric Propulsion Thruster Types

As mentioned in chapter 1, this thesis focuses on micro-Newton propulsion for highly precise AOCS. Considered thruster candidate concepts are RIT, FEEP, or HEMPT. Therefore, in the following subsections provide an overview of the mentioned thruster physics.

2.3.1 Field Emission Electric Propulsion

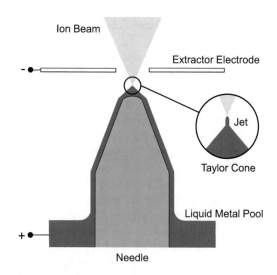

Figure 2.3: The sketch presents the FEEP principle. The needle that is covered by the
liquid metal which is used as propellant lays on a positive potential whereas the
extractor is on a lower potential. Due to the strong electrical field at the needle
tip, called the Taylor cone the propellant is ionised and accelerated [20].

FEEP thrusters, sometimes called emitters, consist of a needle that is either covered
with the propellant, or a capillary inside the needle is filled with the propellant. In
more advanced designs the needle is made of porous tungsten filled with propellant.
Due to the capillary forces the molten propellant is held on the needle tip. Therefore,
no active propellant feeding is required. Liquid metals are used as propellant because
of the high atomic mass, low ionisation potential and good wetting properties. The
metal is stored as a solid body. The propellant becomes liquid via heating. Sketch
2.3 presents the design of a FEEP thruster [20].

By applying a potential difference between the tip of the needle and the extractor
electrode a strong electric field is generated at the sharp needle tip. The field strength
is in the order of 10^{10} V/m and sharpens the liquid to the surface needle into a
Taylor cone, hence the liquid metal is evaporated and directly ionised at this tip. The
electrons created are directly absorbed by the anode tip, whereas the ions created
are accelerated in the electric field between the needle and the extractor electrode.
Therefore, ionisation and acceleration take place in one step and use the same

electric field. To avoid back streaming of electrons from outside into the emitter, the extractor electrode can be negative biased with respect to the spacecraft ground, i.e. the extractor shields the needle and the inner parts of the thruster due to the negative potentials around the extractor. The bias voltage can have values up to -1000 V.

For the strong electrical field that is required for the evaporation and ionisation, the potential difference between the needle and the extractor electrode is typically between 5 kV and 12 kV. In this way, specific impulses up to 8000 s can be realised. The beam current can vary between 1 µA to 100 µA for a single emitter for long term operation. This corresponds to a thrust between 0.1 µN to 12 µN [21, 59].

Dependent on the specific operation parameters, the I_{sp}, the divergence efficiency, the mass efficiency and the PTTR vary. Typical values are an I_{sp} of 3000 s, a divergence efficiency of 0.75 without focusing electrode or 0.85 with focusing electrode, a mass efficiency of 0.40 and a PTTR of 70 W/mN.

In Table 1.2 key parameters of the FEEP are summarised. Due to the required high voltage supply which has to provide up to 12 kV and the required propellant heating, the system complexity is medium compared to the other thruster types (see sections 2.3.2 2.3.3). The lifetime can be limited by the contamination of the extractor electrode with evaporated propellant closes the extraction hole. The lifetime of a FEEP system is strongly dependant on the emitter design. Moreover, the lifetime is limited by the propellant reservoir, i.e. the liquid metal tank has a physical limit because of the propellant feeding via capillary forces.

However, the propellant feeding via capillary force is an advantage too, because no mass flow controller, pressure regulator, tank, or pipework are required. As mentioned, a single emitter is able to work within a thrust range of 120 and with a thrust resolution in the order of 0.1 µN. It was indirectly demonstrated that the FEEP thrusters are able to have a thrust noise below $0.1\ \mu N/\sqrt{Hz}$ in the LISA frequency space, where direct measurements underlined the thrust noise of the thruster, but not in the complete LISA frequency space because of the limitations of the thrust stand used[20, 60].

2.3.2 Radio Frequency Ion Thruster

The RIT is a Gridded Ion Thruster (GIT) that ionises the propellant via a Radio Frequency (RF) electron bombardment. Figure 2.4 shows the basic principle of a RIT. The ionisation takes place inside the thruster inside the so called plasma cavity that is marked as Io in Figure 2.4. The chamber isolates the plasma from the rest of the structure. The axially symmetric discharge chamber is surrounded by the RF-coil. The gas inlet (ins) is on the left side. To accelerate the ions out of the plasma cavity a grid system is placed at the open side of the thruster [61, 62].

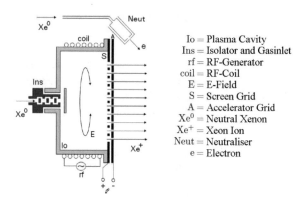

Io =	Plasma Cavity
Ins =	Isolator and Gasinlet
rf =	RF-Generator
coil =	RF-Coil
E =	E-Field
S =	Screen Grid
A =	Accelerator Grid
Xe^0 =	Neutral Xenon
Xe^+ =	Xeon Ion
Neut =	Neutraliser
e =	Electron

Figure 2.4: Simplified view of a two gridded RIT system. The plasma cavity is surrounded by a coil that is used to apply the RF field to the electrons which are inside the plasma cavity to ionise the propellant being injected. The ionised propellant is extracted via a system of grids at the open end of the plasma cavity [61].

To ionise the propellant, an inductive plasma generator is used. The coil around the thruster acts as an antenna that is driven by a RF voltage amplifier. The frequency of the RF depends on the specific thruster geometry and is typically in the MHz regime. Electrons are accelerated inside the RF electric field. With respect to the injected propellant mass flow and therefore to the neutral gas density inside the thruster, the electrons have a specific probability of colliding with a neutral propellant particle, causing the particle to be ionised. If a free electron ionises a neutral particle another free electron is created. The electrons created can then be accelerated again inside the RF field, which leads to a highly efficient ionisation and a high mass utilisation.

The ions created are extracted from the plasma cavity via the grid system. Typically, the grid system consists of two or three grids. Figure 2.4 illustrates an accelerator that consists of two grids. The inner grid (marked as S) is called the screen grid. The screen grid is biased on a high potential in reference to the spacecraft and the neutraliser ground. Due to the high potential of the inner grid, the plasma that is in direct contact with the grid is also raised to the high potential. The potential of the screen grid is typically between 600 V and 1500 V. The second grid is called the accelerator grid. It is biased to a negative potential to prevent electron back streaming to the grid system. The negative potential has to be low enough to avoid the electron back flow as well as to enable an optimal ion trajectory of the extracted ions. Typical acceleration grid voltages are in the range of −70 V to −300 V.

To neutralise the ion beam, a neutraliser is placed in front of the thruster. The neutraliser is shown in figure 2.4, marked as Neut. The neutraliser can be used as an electron source for the thruster ignition as well.

Key parameters of the RIT are summarised in table 1.2. In general RITs have achieved a high maturity level, but due to the requirement of two high voltage power supplies, the Radio Frequency Generator (RFG), a mass flow controller, a pressure regulator, tank, pipework and neutraliser, the complexity of a RIT system is relatively high.

2.3.3 High Efficiency Multistage Plasma Thruster

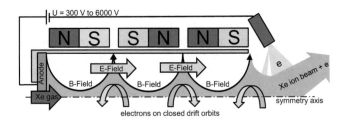

Figure 2.5: Schematic view of the HEMPT concept. The cylindrical discharge channel is surrounded by three magnetic rings, forming, three cusps inside the discharge channel. At the thruster exit an electron source emits electrons to neutralise the extracted ion beam and to maintain the plasma discharge. The electrons are confined inside the cusp field structured magnetic field topology [48, 63].

A typical example for a Direct Current (DC) electron bombardment thruster with an open plasma cavity and hence without a grid system is the HEMPT.

The HEMPT is in development at TED in Ulm. The basic features of the thruster concept are patented by TED [64, 65, 66, 67, 68, 69]. The concept is based on the ideas of Günter Kornfeld et al. to use the specific magnetic cusp field configuration of travelling wave tubes to design a new kind of electric thruster. The thruster is in development since 2000 [24]. The first HEMPT that will be used in space is the HEMPT 3050. It shall be used as the main thruster of the DLR Heinrich Hertz mission [70]. The thruster is able to produce $44\,\text{mN}$ of thrust with an I_{sp} higher than $2400\,\text{s}$.

A schematic of the thruster is given in figure 2.5. It depicts a cut through the cylindrical HEMPT. The dielectric discharge chamber (in grey) is surrounded by a system of periodically poled permanent magnets. This magnet stack forms the special magnetic field topology which separates the discharge chamber into several magnetic cells. This configuration is also used in travelling wave tubes. The magnetic field lines are represented as black arrows. The anode (in dark red) is mounted at the inner side of the thruster. The anode is the high voltage connection as well as the gas inlet. In figure 2.5 the anode is placed on the left side (upstream side). At the thruster exit, also called the downstream side, a neutraliser, sometimes called the cathode, is placed. Between the anode and the cathode the potential difference ranges from 300 V to 6000 V.

From the neutraliser electrons are pulled towards the thruster, due to the electric field between the anode and the cathode. The electric field configuration is illustrated as bright red arrows in the schematic. Without neutral gas background and hence without collisions, the motion of the electrons from the downstream side to the upstream side of the thruster is hindered by the radial magnetic field topology and can be described by equation 2.13. Because of the magnetic field topology, the electrons are trapped in front of the thruster exit and each internal cusp due to the magnetic mirror effect (as described in 2.2). At the exit cusp the trapped electrons define a space charge potential close to the cathode potential which acts similar to an acceleration grid in a GIT. The electron confinement is almost perfect if no neutral gas background is present. Accordingly, the impedance of the thruster is almost infinite. With activated propellant flow the neutral gas expands into the discharge chamber, setting up a specific neutral gas density distribution.

The possibility of electron-neutral collisions are dependent on the neutral gas pressure distribution inside the discharge chamber. Elastic electron neutral collisions lead to electron transport through the thruster. While gaining kinetic energy in the E-field, electrons are pushed from the B-field lines where they are trapped onto other B-field lines. The electrons move inside of the thruster and the impedance of the thruster decreases.

Inelastic collisions can ionise neutral particles, as described in section 2.2, creating a positive ion and two electrons. The electrons continue their motion dependent on the E-field and B-field inside the thruster and their residual temperature. Generally, electrons can collide several times until their kinetic energy, and therefore their velocity, is too low to get trapped inside the B-field and they becomes thermalised. The thermalised electrons are pulled through the E-Field to the anode, closing the electric circuit between anode and cathode.

The heavy propellant ions are typically not affected by the B-Field due to the dimensions of a HEMPT (with discharge chamber diameters between 3 mm and 200 mm) and the mass and the energy of the propellant ions. This fact can be illustrated via equation 2.11 which shows that the typical Larmor radius of a xenon ion inside the thruster is in the order of thousands of millimetres. Thus the thruster diameter is much smaller than the Larmor radii of the ions.

However, the ions are accelerated towards the thruster exit by the electric field. Moreover, the electrons are focused on the rotation axis because of the magnetic field topology and therefore the ions are also focused on the rotation axis.

Additionally, some ions collide with the walls of the discharge chamber which leads to a positive charging of the non conductive walls [71, 72]. The positive charged walls are not neutralised by electrons because the electrons are confined inside the B-Field. The positive charge of the walls leads to a better confinement of the ion beam inside the thruster, with almost no interaction of the plasma and the discharge chamber walls.

As presented in figure 2.2, the electrons gyrate along the magnetic field lines and are reflected inside the cusps. Therefore the ionisation of the neutrals takes place inside the whole discharge chamber and is only dependant on the density of the propellant particles. Because of the continuous ionisation inside the thruster, the plasma potential is almost flat inside the discharge chamber. Due to the B-field geometry, as well as the low neutral gas density at the exit cusp, a negative space charge is created in front of the thruster. Therefore, the plasma potential drops down to zero inside a relatively short distance. Figure 2.6 illustrates the qualitative trend of the plasma potential inside the thruster. Ideally, the small slope of the plasma potential pushes the ions to the thruster exit.

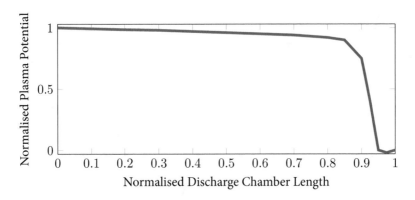

Figure 2.6: The curve illustrates the qualitative plasma potential characteristic inside the plasma cavity of a HEMPT. The plasma potential inside the cavity has only a small slope. At the thruster exit the potential drops rapidly to ground potential. Thus, the ion acceleration almost exclusively takes place at the thruster exit [71].

Compared with other electric thruster technologies the ionisation zone is relatively wide, which explains why observations show that almost 20 % of the ions produced are double ionised [57].

The HEMPT requires typically one high voltage supply, a mass flow controller, a pressure regulator, a tank, pipework and a neutraliser. Of particular importance, the design of the required Power Processing Unit (PPU) can be rather simple. Consequently, the complexity of a HEMPT at system level is low. The absence of a grid assembly and the long ionisation area leads to a high dynamic range of the thruster. Due to the excellent ion confinement, no lifetime limiting erosion of the ceramic discharge chamber takes place. Up to now all life limiting processes are found to be negligible.

2.4 Thruster Characterisation

Before an electric thruster can be used on a satellite a lot of development and testing activities have to be performed to prove that the thruster is able to fulfil all requirements, like every other major and minor spacecraft subsystem. The most important key parameters, like the PTTR, the specific impulse (I_{sp}) and the thrust were introduced in section 2.1. But, dependent on the purpose of the thruster, also other parameters play an important role such as the thrust noise of the thruster which is the most important key figure for highly precise AOCS thrusters.

Direct or indirect measurement techniques can be used to determine parameters of electric thrusters. For example, to determine the thrust, a direct thrust measurement can be performed, or the measurement of the beam power via a plasma diagnostic setup followed by the derivation of the thrust (see equation 2.5) can be used. In general, for the development and space qualification of thrusters a detailed characterisation with various devices has to be performed. The use of different measurement techniques is required to cross check the data obtained to reduce the measurement uncertainties and to avoid misinterpretations. However, to be sure that the thruster generates the predicted thrust value, a direct thrust measurement has to be performed with each thruster. If the thrust noise plays an important role this parameter also has to be directly verified.

Over the years, a lot of different diagnostic methods for electric thrusters were developed. The most simple approach is to measure the anode voltage, the anode current and the mass flow that can be transformed to the thrust, the PTTR and the I_{sp} [17]. Unfortunately, losses are not taken into account, therefore unpredictable measurement errors can occur. To take all errors into account, a direct thrust measurement is the preferred measurement technique. When using a thrust balance the parameters mentioned can be very precise determined. To determine other parameters such as the divergence efficiency, electrical efficiency, etc., measurements of the plasma plume can be performed. Typically, electrostatic probes are used for this purpose. Due to the complexity of plasma physics almost every parameter is important for EP development, characterisation and qualification. Therefore, various instruments should be used.

The characterisation of micro-Newton electric propulsion is especially challenging, due to the small forces and currents. In this section the basics of micro-Newton direct thrust measurement and an introduction of plasma plume characterisation will be given.

2.4.1 Direct Thrust Measurement

In general, there are several ways to measure forces in the micro-Newton regime including weighing scales, piezo micro balance, or pendulum balances. However, because of the typical mass to thrust ratio of electric thrusters pendulum balances are preferred for testing. The typical mass to thrust ratio of an electric thruster is in the range of 10^7 kg/N to 10^8 kg/N. State of the art piezo micro balances are able to measure within a mass to thrust ratio up to 10^6 kg/N, whereas pendulum balances are able to measure extremely high mass to thrust ratios [73]. Currently, two kinds of pendulum balances are used for micro-Newton thruster testing, vertical pendulum balances (also called hanging balances) and horizontal pendulum balances, (sometimes called torsion pendulum balances, or torsion balances). The primary difference of the pendulum types is how the gravity force acts on the pendulum arm. Gravity has no influence on an ideal torsion balance because the gravity vector is perpendicular to the plane of motion. Therefore, the torsion coefficient (κ in Nm) is independent of the gravity vector. However, for vertical pendulums gravity acts as a restoring force . Thus, κ consists of the torsion coefficient from the pendulum bearing (κ_s) and the torsion coefficient caused by gravity, with m_p being the pendulum mass and l_{cog} the distance between the Centre of Gravity (CoG) and the pendulum bearing:

$$\kappa = \kappa_s + m_p \cdot g \cdot l_{cog} \quad . \tag{2.17}$$

Both balance types follows the same measurement principle, only the orientation of the pendulum is different. Pendulum balances measure force by the measurement of the pendulum deflection (Θ) that can be multiplied with the pendulum torsion coefficient to determine the applied thrust (F). It can be written as

$$M = \kappa \cdot \Theta \quad , \tag{2.18}$$

$$F \cdot l = \kappa \cdot \Theta \quad , \tag{2.19}$$

where l is the pendulum length and M is the applied momentum. For small deflections, the thrust can be determined by a simple translation measurement:

$$F = \frac{\kappa \cdot \Theta}{l} = \kappa \cdot \frac{\Delta x}{l^2} = K\Delta x \quad , \tag{2.20}$$

where K (in N/m) is now the spring rate or spring coefficient and Δx (in m) is the measured translation.

The pendulum can be described as a physical pendulum, i.e. the second order equation of motion can be written as

$$I\ddot{\Theta} + c\dot{\Theta} + \kappa\Theta = F(t)l \quad , \tag{2.21}$$

where c is the damping coefficient, I is moment of inertia, κ is the torsion coefficient and $F(t)$ is an applied force acting at the distance l from the pendulum rotation point. L,I,c and κ are assumed to be constants. Equation 2.21 can be written in the standard form

$$\ddot{\Theta} + 2 \cdot \zeta \cdot \omega \cdot \dot{\Theta} + \omega_n^2 \cdot \Theta = F(t) \cdot \frac{l}{I} \quad , \tag{2.22}$$

where ω_n is the eigenfrequency, or natural frequency of the undamped pendulum,

$$\omega_n = \sqrt{\frac{\kappa}{I}} \quad , \tag{2.23}$$

and ζ is the damping coefficient,

$$\zeta = \frac{c}{2} \cdot \sqrt{\frac{1}{I \cdot \kappa}} \quad . \tag{2.24}$$

The equation can be solved to calculate the response of the balance to an arbitrary input force. Figure 2.7 illustrates the different special cases that are relevant for a thrust balance. On the y-axis the normalised response is plotted versus time in natural periods. The main difference between the curves shown is the damping factor. The time required to settle within 2 % of the steady state deflection is called settling time. The dotted curve presents the behaviour of an underdamped system to reach steady state starting from a initial deflection. Due to the low damping factor it needs more than 3 natural periods to settle. The solid curve presents the critically damped case, where the settling time is minimised and no overshooting occurs. The behaviour for higher damping factors is illustrated with the dash-dotted curve. It is clear that the higher damping factors leads to a higher settling time.

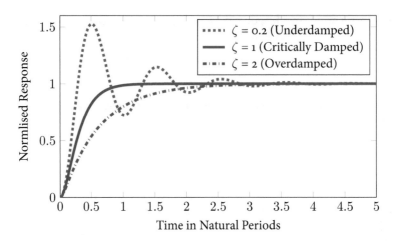

Figure 2.7: Illustration of the typical step response of pendulum balances with different damping coefficients. The dotted curve represents the underdamped case with low response time and overshooting. The solid line illustrates the critically damped case with a longer response time and no overshooting. The dash-dotted curve shows the overdamped case with the longest response time and also no overshooting.

The specific behaviour of a pendulum balance can also be expressed using the Bode plot of the system [74], presented in figure 2.8. The upper figure, the magnitude plot, shows the logarithmically plotted magnitude versus the logarithmically plotted and normalised frequency. Whereas, the lower figure, the phase plot, illustrates the phase angle in radians against the normalised, logarithmically plotted frequency. The dotted curves present the underdamped case, the solid lines show the critically damped case and the dash-dotted curves present the overdamped case. In the underdamped case an amplification at the first eigenfrequency is visible. At higher frequencies the balance acts as a low pass filter. At the phase plot the dotted curve presents a phase shift of $\frac{1}{2}\pi$. The solid and the dash-dotted curve present the thrust stand behaviours with higher damping factors. In general, the pendulums are acts like low pass filters. With rising damping factors the cut-off frequency shifts to lower frequencies and the slope of the curves in the phase plot decreases. In any case, for frequencies above the eigenfrequency the response is attenuated. Hence, the sensitivity of a pendulum thrust stand varies with the frequency.

In general, the overall performance of thrust balances can be described by several metrics such as, sensitivity, resolution, accuracy, measurement range, measurement bandwidth and long term stability [73]. These key figures can be traded versus each

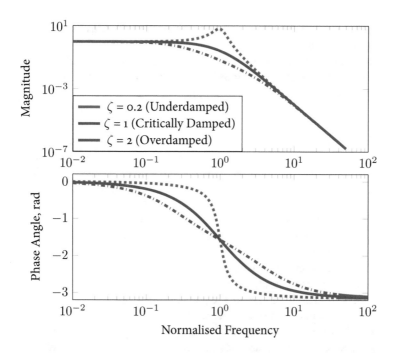

Figure 2.8: Bode plot of an arbitrary pendulum balance. The dotted curves illustrate the underdamped case, the solid curves present the critically damped case and the dash-dotted curves show the overdamped case. The upper figure is the magnitude plot and the bottom figure is the phase plot. The Bode plot allows a simple assessment of damping and phase variations for different frequencies.

other in order achieve an optimal balance behaviour with respect to the targeted application.

The sensitivity (S_{cal}) can be defined as the change of the output value of a measurement device in relation to the variation of the input value that produces the change. Therefore, the sensitivity of pendulum thrust stands have the unit rad/N. For deflections where the small angle approximation is valid, the sensitivity can be given in N/m with respect to distance of the measurement point from the pivot. In this specific case the units of the sensitivity and the spring constant are the same. Typically, the real sensitivity of the thrust stand has to be estimated via a calibration. The sensitivity depends on the ability to measure the deflection or translation of the pendulum. As noted above, the sensitivity can vary with frequency. This is especially important for noise measurements, or measurement of dynamic thrust loads. Hence, to perform noise measurements above the eigenfrequency, the sensitivity must be corrected with the transfer function.

The resolution is defined as the smallest possible difference between two thrust inputs. In practice, the resolution is limited by the noise floor of the thrust measurement. Therefore, to achieve a high resolution, a minimal noise floor is necessary. Typical noise sources are mechanical noise due to vibrations in the environment or electrical noise that is induced by the electronics used or other nearby electronics.

Periodic changes of temperature can directly couple into the mechanical structure which is similar to long term mechanical vibrations. Moreover, temperature changes can also influence the electronics used. Usually, temperature variations limit the resolution at low frequencies. The ability to measure at low frequencies is called stability.

Additionally, the measurement bandwidth defines the achievable resolution in a specific frequency spectrum. The measurement range presents the minimal and the maximum measurable thrust.

The error between the measurement value of the thrust stand and the real thrust value can be described as accuracy. High accuracy of a thrust balances requires a minimising of systematic errors and an adequate calibration.

As mentioned, torsion balances and vertical pendulum balances can be used for micro-Newton thrust measurements. Both concepts have specific advantages and disadvantages and it has been demonstrated that both concepts are suitable for thrust measurements [75, 47, 38, 76, 77, 44]. A simplified version of the balances are presented in figure 2.9 to illustrate the general balance design and the basic components. Figure 2.9 (a) shows a torsional balance. It consists of a bearing that is typically a torsion spring, a damper, a counterweight to keep the CoG in the axis of the torsion spring, a calibration device, a translation readout and the thruster which shall be tested. In principle, vertical pendulum balances consist of almost the same parts. However, usually leaf springs are used to bear the pendulum and a counter balance is not required. A simplified vertical pendulum balance is presented in figure 2.9 (b).

To determine the position of the pendulum arm, different types of translation, angle or position sensors can be used. Sensors that have been used are capacitive sensors, linear variable differential transformers, optical reflection ratio-based displacement transducers, or optical cavities [44, 38, 73, 78]. Naturally, the different kinds of translation/deflection readouts have different advantages and disadvantages. The general rule of thumb is that with increasing resolution the complexity of the readout system increases. For example, a simple capacitive sensor consists only of a plate capacitor and simple electronics which measure the variation of the system capacity, whereas an optical readout like a heterodyne interferometer would require a laser, different optics and a much more advanced electronic system [8]. However,

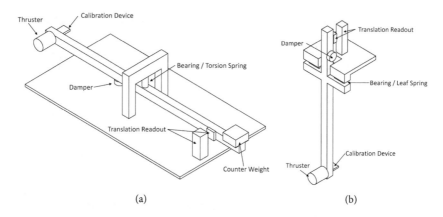

Figure 2.9: The sketches illustrate the most simple mechanical configuration of a torsion balance (a) and of a pendulum balance (b). Both types require a bearing, a calibration device and translation sensor. Additionally a damper can also be part of the setups.

a laser interferometer has typically a resolution which is some orders of magnitude higher than a capacitive sensor. Thus, dependent on the targeted resolution, the use of a more complex readout could be advantageous.

The spring rate of the balance, the readout and the specific distances between the thruster, the pivot and the deflection/translation sensor define the sensitivity of the thrust stand. Therefore, a higher resolution of the translation readout leads to a higher sensitivity. Moreover, the readout should be linear and as stable as possible.

To achieve a sufficient balance resolution, accuracy and low measurement errors, a reliable and trustworthy calibration is necessary. Common calibration methods are the use of calibration weights, gas dynamically calibration or electrostatic force calibration [79].

Dependent on the chosen measurement procedure, a suppression of the eigenfrequency of the pendulum is required to achieve the highest possible balance resolution. Therefore, a damper is required as presented in figure 2.9. The damping of the system should have an almost ideal characteristic, such as a frictionless operation to enable good repeatability. Eddy current brakes are typically used as passive and contactless damping elements [38, 73].

The general features of the two concepts are summarised in table 2.1. The torsion balance is insensitive to ground noise because gravity has almost no influence on the horizontal pendulum. However, the horizontal pendulum arm always requires a minimum of space and therefore the implementation of a miniaturised torsion balance is challenging. The insensitivity to gravity make the use of tuning weights to adapt the spring rate impossible.

The vertical pendulum balance is sensitive to ground noise because gravity acts directly on the balance. However, the spring rate can easily be varied with tuning weights which leads to a wider measurement range. The vertical implementation can also be more easily miniaturised. The long term stability of both concepts is coupled to thermomechanical noise and therefore directly dependent on temperature variations.

Table 2.1: Summary of a qualitatively tradeoff between the two considered thrust balance types.

Feature	Torsional Balance	Vertical Pendulum
Typical bearing type	Torsional spring	Leaf spring
Required system space	Medium	Low to medium
Ground noise sensitivity	Low	High
Measurement range	Fixed	Flexible
Stability	Medium	Medium

In summary, the direct thrust measurement is the most accurate way to quantify important key figures of an electric thruster. Typically, pendulum balances in different variations are used. But especially the long term stability of a single pendulum is limited due to thermomechanical noise.

2.4.2 Indirect Thrust Measurement

The second way to assess the key figures of an electric propulsion device is to analyse the plume of an ion source, which requires surface integration and/or plume symmetry assumptions. Plasma plume diagnostic tools are typically used to evaluate the thruster and its performance. It is common practice to measure the energy content and the spatial current density distribution. According to section 2.1, the beam power, the beam divergence and of course thrust can be determined. Which is way, it is called indirect thrust measurement. Moreover the specific impulse and the PTTR can be obtained.

In the last decades various set of plume diagnostic tools were developed. Typical examples are Langmuir probes, Faraday probes or cups, RPA, ExB probes, mass spectroscopy, LIF, infrared imaging, neutral particle flux probes and several other probes [80, 34, 81, 25, 42, 82, 83].

The indirect thrust measurement is based on the assessment of the beam current (I_b) and the acceleration voltage of the beam (V_b). The acceleration voltage corresponds to the energy of the ion beam for single charged ions. In practice, the distribution of the ion beam is not uniform, therefore the beam divergence has to be considered. To measure the beam current density and distribution, Faraday probes are commonly used. While energy selective detectors have to be used to determine the beam energy, for example RPAs, mass spectrometers, or LIF measurements.

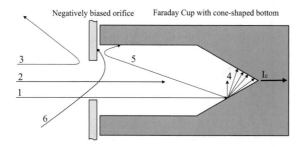

Figure 2.10: Schematic of a Faraday Cup used to measure the ion current density of an electric thruster. The cup (in dark grey) collects the ions, while the electron shield (in light grey) shields the collector from electron impingement. The vectors illustrate the behaviour of different kinds of charge particles; 1: Impacting ion which produces secondary electron emission; 2: Impacting ion; 3: Deflected electron; 4: Secondary electrons that cannot leave the cup due to the conical cup bottom; 5: Secondary electron that cannot leave the cup because it is reflected back by the electron shield; 6: Ion with a large incidence angle that is deflected by the electron shield; I_c: measured ion current.

A standard implementation of the Faraday probe is the Faraday cup. A sketch of the Faraday cup is presented in figure 2.10. In the basic configuration shown it consists of a cup that collects the ions (trajectory 1 and 2) and a negatively biased orifice that shields the cup from the electrons which are part of the ion beam, presented as trajectory 3. The end of the cup is conical to prevent secondary electrons escaping, illustrated as trajectory 4 and 5. The E-field topology created by the negative biased orifice also deflects secondary electrons back to the cup. The conservation of the electrons created is important for a precise ion beam current measurement because the systematically loss of charges would lead to higher measurement uncertainties. The simple design is partly able to suppress slow ions produced outside of the thruster (trajectory 6). A rotatable array of Faraday cups is usually arranged around the thruster under investigation to map the whole plume, allowing the current that is measured at the cups to be integrated to obtain the beam current.

A representative example of a plume characterisation via a Faraday cup assembly is presented in figure 2.11, where the current density (in A/m^2) is plotted versus the angle (in deg). The measurement was performed by A. Keller at the University of Gießen [32]. The three curves illustrate the shape of the plasma plume of the thruster variation tested. In the presented case the housing material was varied.

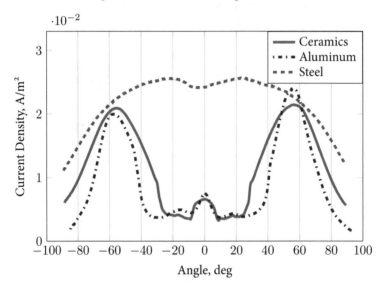

Figure 2.11: Example of a typical Faraday cup measurement. The plot presents three different measurements of three different thrusters. The curves illustrate the diversity
of plume shapes which can be observed dependent on specific geometry of the
thruster tested [32].

To measure the energy of the ion beam, RPAs have become the standard instrument [82, 81, 37]. The sensor measures the current on a defined collector surface due
to the impacting ion beam. In contrast to Faraday probes, the ion beam has to flow
through a positive potential difference, decelerating the particles. The measured
beam current depends on the energy of the particles and the potential difference, i.e.
the specific ion energy can be determined by increasing the positive potential until
no current can be measured.

Typically, the potential difference of the detector is applied by a system of dif-
ferently charged grids. For example in a three grid RPA the first grid would be on
fixed negative potential to repel electrons, similar to the electron shield orifice in
the Faraday cup design. The second grid would have a variable positive potential to
apply the retarding potential. And, the third grid would be on a negative potential

to repel electrons created by secondary electron emission due to the impact of ions on the collector. The RPA should also be placed on a rotatable jib arm to quantify the ion energy distribution over all angles.

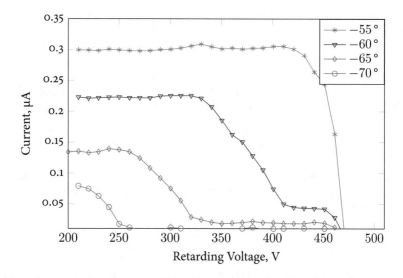

Figure 2.12: The curves shows a representative RPA measurement, where the measured current is plotted versus the applied voltage. The curves illustrate measurements at different angular positions inside the plume of the thruster tested. It is shown that the ion energy varies dependent on the plume angle [32].

Figure 2.12 presents a typical RPA measurement that is part of the measurements which were created by A. Keller at the University of Gießen [32]. The current measured is plotted versus the retarding potential (in Volts). At the time when the measurement was made, the thruster operated at an anode voltage of 500 V. Each curve illustrates a measurement at a specific angular position of the RPA inside the plume. It can be seen, that with rising retarding voltage the current measured decreases and above 475 V the current measured is zero. Via numerical integrating of the curves the acceleration efficiency can be determined. At minus 55 ° the beam current and the acceleration efficiency are at their maximum. Whereas, the acceleration efficiency and the beam current are at their minimum at minus 70 ° (in accordance to figure 2.11).

To calculate the thrust via plasma diagnostic measurements, the beam current and the acceleration efficiency at all angles must be considered. The total current in a specific segment of the sphere in front of the thruster can be calculated by

$$I_{total,\varphi} = \rho_\varphi \cdot 2 \cdot \pi R^2 \cdot (cos(\varphi - \Delta\varphi/2) - cos(\varphi + \Delta\varphi/2))/2 \qquad , \quad (2.25)$$

where ρ_φ is the current density at the specific angular position, R is the radius of the sphere and φ the angular position of the specific segment.

Hence, the overall ion beam current can be estimated by the sum of the particle currents at all angles ($I_b = \sum_\varphi I_{total,\varphi}$).

As mentioned, the angular ion current distribution inside the plume produces an impulse loss which can be illustrated by the divergence efficiency. Therefore, in case of a two dimensional symmetric beam, the divergence efficiency can be written as

$$\gamma = \frac{\sum_\varphi I_{total,\varphi} \cdot cos\,\varphi}{I_b} \qquad . \qquad (2.26)$$

In the scope of this thesis also thrusters with significant amount of double charged ions, such as HEMPTs, are discussed [71]. Hence, the divergence efficiency must be corrected with the thrust correction factor α, which can be written as

$$\alpha = \frac{I^+ + \frac{1}{\sqrt{2}} \cdot I^{++}}{I^+ + I^{++}} \qquad , \qquad (2.27)$$

where I^+/I^{++} is the fraction of the double ion current in the beam [2]. Henceforth, the divergence efficiency can be redefined as

$$\gamma = \alpha \cdot \gamma \qquad . \qquad (2.28)$$

In addition to γ, the mass efficiency η_m can also be derived from the ion beam current measured and the ratio between double and single charged ions as

$$\eta_m = \frac{1 + \frac{1}{2}\frac{I^{++}}{I^+}}{1 + \frac{I^{++}}{I^+}} \cdot \frac{I_b}{q \cdot \dot{m}} \qquad . \qquad (2.29)$$

To complete the efficiency parameter determination of the thruster, the acceleration efficiency can be measured via a RPA measurement. Figure 2.12 illustrates a typical measurement result where the measured ion current is plotted versus the retarding voltage. The plot shows that the ion energy varies over different angular measurement positions and therefore the acceleration efficiency varies across the plume, in addition and independent to the variation of the ion current. Thus, the acceleration efficiency can be calculated by

$$\eta_v = \frac{\sum_\varphi I_{total,\varphi} \cdot U_{b,\varphi}}{I_{total,\varphi} \cdot U_a} \qquad , \qquad (2.30)$$

where $U_{b,\varphi}$ is the beam potential at the specific measurement position and U_a is the measured anode potential of the thruster.

The thrust can be determined with the efficiency and the thruster parameter measured such as anode potential and propellant mass flow as

$$F_{t_{in}} = \gamma \cdot \eta_m \cdot \dot{m} \cdot \sqrt{\frac{2 \cdot q \cdot \eta_v \cdot U_a}{m_{ion}}} \qquad . \qquad (2.31)$$

It is also possible to estimate the specific impulse with equation 2.1 and the PTTR with equation 2.7.

The above shows, that indirect thrust measurement allows ascertaining the major thruster parameters and the determination of the most important loss terms of the thruster. In contrast to the direct thrust measurement, different detectors and parameters are required for the indirect thrust measurements and the physics behind the measurement principles are complex. Thus, the overall uncertainty of indirect thrust measurement is typically higher. However, due to the determination of several other thruster parameters, indirect thrust measurement is of special importance for electric thruster development and characterisation.

3 Micro-Newton Thruster Test Facility

In line with the requirements (table 1.3) a micro-Newton thruster test facility has been developed. The facility consists of a vacuum tank, a highly precise thrust balance, a plasma diagnostics setup and all other necessary hardware to perform EP development, testing and characterisation. As introduced in chapter 1, section 1.4, the micro-Newton thruster test facility developed can be used for the complete characterisation of AOCS thruster candidates for future scientific space missions such as LISA. A detailed explanation of the major components of the facility will be provided in the following sections.

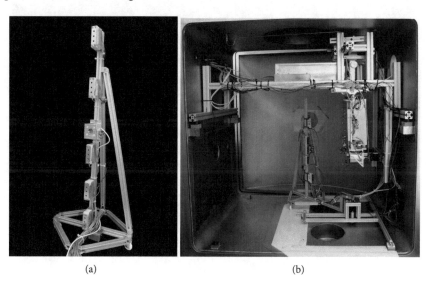

(a)   (b)

Figure 3.1: In figure (a) the plasma diagnostic assembly is shown. All instruments are mounted on a rotatable jib arm made of Item profiles. Figure (b) presents the inner view of the vacuum tank to provide an overview of the complete measurement setup. At the top part the micro-Newton thrust balance is visible, with the fully integrated RITμX assembly. The jib arm can be seen in the middle of the picture.

© Springer Fachmedien Wiesbaden GmbH 2018
F. G. Hey, *Micro Newton Thruster Development*,
https://doi.org/10.1007/978-3-658-21209-4_3

Figure 3.1 presents an overview of the instruments developed which are part of the micro-Newton thruster test facility. On the left side, in figure a) the plasma diagnostic is shown. It consists of a set of Faraday cups and one RPA. All instruments are mounted on a rotatable jib arm which allows a 180° plasma plume analysis. The arm is also visible in the middle of figure b) which shows the inner view of the vacuum tank. On the top of figure b) the micro-Newton thrust balance is shown. At the end of the thrust balance arm (in the middle of the picture, right side) a RITμX is shown as thruster under test. As mentioned in chapter 1, the instruments developed enable a complete experimental characterisation of possible micro-Newton thruster candidates for the targeted missions such as LISA and other future scientific and Earth observation space missions.

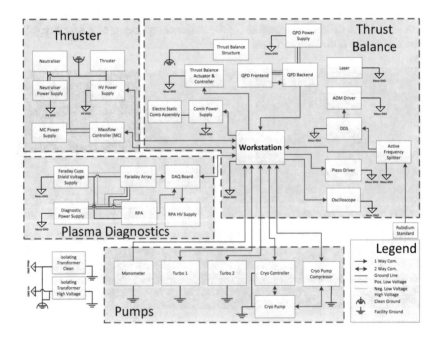

Figure 3.2: Simplified overview of the electrical connections between the sub assemblies of the micro-Newton thruster test facility. The facility consist of four main sub assemblies, which are controlled by the workstation. The data connections are shown as black arrows, the DC voltage is shown as red and blue lines and the DC high voltage as yellow lines. The ground connection of each equipment is also presented.

Figure 3.2 provides a simplified overview of the major sub assemblies and their interconnections. The four sub assemblies are the vacuum pumps and infrastructure, a highly precise and highly stable micro-Newton thrust balance, a plasma diagnostic setup and the thruster under test. The diagram provides information about how these major parts are grounded and how communication lines are connected.

An x86 workstation functions as the central control unit. The workstation provides several interfaces to all other instruments and can be used to control the measurement hardware used. Two FPGA-boards are integrated inside the workstation which provide digital real time controlling of actuators and other equipment. Due to the FPGA-boards, the workstation is an incremental part of the thrust balance. The workstation is also used as graphical interface to the thruster and facility operator and it stores the complete measured data.

A grounding concept with three different grounds is implemented in order to shield the sensitive measurement hardware from electrostatic discharge potentially caused by the electric thruster. The ground lines are called measurement ground, High Voltage (HV) ground and facility ground. The measurement ground and the high voltage ground are merged at a star point that is directly connected to a ground rod, also called clear ground. As star point a copper plate at the vacuum tank is used, this implies that the whole vacuum chamber can be used as ground. To isolate the measurement ground and the HV ground from the facility ground, two isolating transformers are used. Due to the defined grounding of all instruments and the direct connection of all devices to the clean ground an electrically quiet measurement environment was established. Additionally, in the unlikely event of an arcing no leakage current flows through the several instruments in order to avoid damages of the instruments.

3.1 Vacuum Facility

The most generic part of an EP test facility is the vacuum system, because the testing of almost every electric thruster requires a sufficiently low background pressure to guarantee the proper operation of the thruster and to generate reliable measurement results. Thus, the quality of the vacuum has to be similar to the space environment [55]. The vacuum tank used has to be large enough to avoid negative influences from the chamber wall to the thruster and it has to house all required measurement equipment, i.e. the vacuum chamber should be as large as possible. However, a large chambers needs more time to be pumped out, or higher pumping speeds. Larger tanks are also more expensive in purchasing and servicing. Additionally, other characteristics like handling, accessibility, number of flanges, etc. plays an important role.

A tradeoff between the mentioned boundary conditions had led to a rectangular 1500 l vacuum tank. The custom designed vacuum tank has been manufactured and leakage tested by Just Vacuum. Table 3.1 summarises the most important parameters. The inside of the chamber is 880 mm deep (without doors), 1200 mm wide and 1200 mm high. Two doors, one on each flank, enable easy access to every instrument which is inside the chamber. More than 20 flanges allow the installation of various pumps, instruments and interconnections. At the inside of the chamber a set of pre-installed mounting points enables a simple and adaptable placing of different measurement devices. The seals of the doors are custom Viton seals, whereas all other flanges are ISO-KF, ISO-K standard flanges. Thus, all seals are made of Viton. The elastomer sealings are enabling cost efficient and fast handling of the chamber between the measurements and during setup modifications. The tested leakage rate of the chamber is smaller than 10^{-9} mbar l/s.

Table 3.1: Summary of the key parameters of the vacuum facility which consists of the vacuum vessel, the installed pumps, a manometer, various vacuum feedthroughs and valves and the required tubing.

Part/Feature	Size	Quantity
Body	880x1200x1200 mm	1
Doors	250x1200x1200 mm	2
Various flanges	from ISO-KF 16 to ISO-KF 500	23
Seal material	Viton	
Tested leakage Rate	10^{-9} mbar l/s	
Installed pumping speed	11400 l/s	
Minimal measured pressure	$1.2 \cdot 10^{-7}$ mbar	
Operational pressure (0.5 sccm ballast)	$3.5 \cdot 10^{-6}$ mbar	

The seismic noise background of the facility plays an important role for micro-Newton propulsion testing. It has been demonstrated that especially for micro-Newton thrust measurements or other highly precise measurements, the decoupling of the facility from the ground is necessary [48, 8]. Therefore, the whole vacuum tank is pivoted on isolator feeds. The feeds are Commercial Off The Shelf (COTS) pneumatic vibration isolators from Newport [84]. The connection between the isolators and the tank is realised with a structure made of Item aluminium Profile 8 standard bars. The whole setup is presented in figure 3.3. Due to the Item profile structure, the setup can be easily adapted if required.

Dependent on the thruster type which shall be tested, different background pressure levels had been established as standard, e.g. Hall Effect thrusters are usually tested in the 10^{-5} mbar regime [85], whereas GITs are tested in the 10^{-6} mbar regime [86]. Therefore, the general pressure level during thruster testing should be in 10^{-6} mbar regime to reduce possible side effects and to be able to test almost all micro thruster types. To roughly estimate the required pumping speed, the gas ballast which is generated by the thruster can be used, since other gas ballasts are typically more than two or more orders of magnitude lower. The propellant flow rate of micro electric thruster is usually below 0.5 sccm [87, 32]. Thus, according to the ideal gas equation a pumping speed in the 10^4 l/s range is required, therefore a pumping system with several stages is necessary. Dependent on the required speed different pump technologies are available, like turbo molecular pumps, ion getter pumps, membrane pumps, rotary vane pumps or cryo pumps [88].

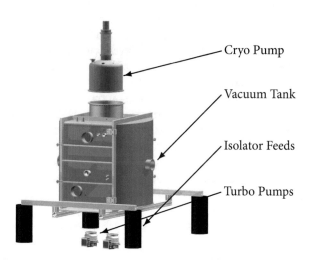

Cryo Pump

Vacuum Tank

Isolator Feeds

Turbo Pumps

Figure 3.3: Illustration of the vacuum facility that presents the main components of the setup. They are the cyro pump, the vacuum tank and the turbo pumps. All parts are mounted on an Item structure that is born by isolator feeds to decouple the whole assembly from the ground in order to reduce ground noise impact.

To fulfil the mentioned requirements, a pump system that consists of three stages has been selected. As first stage, or sometimes called fore stage, a 5.5 l/s rotary vane pumps is installed. Only with the rotary vane pump the vacuum tank can be evacuated to the pressure levels below 10^{-1} mbar. Rotary vane pumps are relatively loud and noisy. Hence, the pump is placed in a separated container outside the laboratory. A 10 m corrugated tube is used as connection to the next pumping stage. The tube is also decoupling the vibrations from the fore stage pump to the facility. The second pumping stage is formed by two 700 l/s Pfeiffer High Pace 700 turbo pumps. The pumps are shown in figure 3.3. To decouple the turbo pumps from the vacuum tank, special damping elements (Pfeiffer Vibration damper for HiPace) are used. The turbo pumps are able to evacuate the chamber down to 10^{-6} mbar without gas ballast.

Higher pumping speeds are required to reach the targeted pressure level of 10^{-6} mbar, with active gas ballast. Different pump technologies are offering sufficient pumping speeds in the order of 10000 l/s such as oil diffusion pumps, cryo pumps, or a set of big turbo pumps.

In EP testing cryo pumps had become standard because of the disadvantages of the other pump systems and the unique advantages of the cryo pumps. The disadvantage of oil diffusion pumps are the required oil and hence the risk of oil contamination. The use of a high number of turbo pumps is not optimal as well, according to the enormous investment costs of such an assembly. In contrast, cryo pumps offer high pumping speeds, paired with clean vacuum environment and an easy handling. Additionally, cryo pumps are available as COTS and a lot of experience exists with EP testing and cryo pumps.

The cryo coolers which are available on the market use the Gifford/McMahon principle. Gaseous helium is compressed up to 30 bar within a compressor. The compressed helium is transported to the cryo head where it is expanded in a displacer. The expansion is isothermal, thus heat is taken up from the cold plate of the cryo head i.e. the plate is cooled. The expanded helium is transported back to the compressor. The process starts from the beginning. The process can be implemented as a closed cycle, thus no helium is lost during operation. This is saving costs and maintenance effort.

Figure 3.4 provides an overview of the whole vacuum setup and summarises the previously described vacuum setup. The compressor of the cryo pump and the forestage pump are placed outside the laboratory. Beside the notable reduction of the sound level inside the laboratory, the use of the external rack reduces the seismic noise floor of the instruments inside the chamber significantly. A rotary slide valve between the forestage pump and the turbo pumps allows a separation of the pump and the forestage vacuum and it is used to avoid oil back streaming during venting. A second valve is mounted to the chamber for the venting of the chamber with dry nitrogen.

To measure the pressure inside the vacuum vessel, a manometer is part of the facility. The Pfeiffer Pirani/Bayard-Alpert manometer combines two measurement

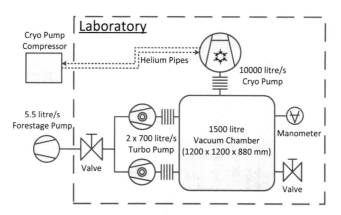

Figure 3.4: Flow chart of the vacuum setup. The pumps are decoupled by membrane bellows from the vacuum chamber. The cryo compressor and the forestage pump are placed outside of the laboratory to reduce the seismic noise level in the inside of the laboratory.

principles in a single housing. In the low pressure regime the Pirani element measures the pressure inside the tank. Whereas, beginning below 10^{-3} mbar the hot cathode gauge also called Bayard Alpert manometer takes over. The manometer provides an accurate pressure level determination also for different gas combinations inside the vacuum chamber down to $5 \cdot 10^{-10}$ mbar. The pressure level can be digitally accessed via a RS-232 interface.

The major disadvantage of cryo coolers for highly precise metrology is the seismic noise caused by the expansion of the helium. To overcome this problem, Gerd Jacob from ESO performed a detailed analysis of COTS cryo coolers. The survey comprehend six different coolers, measurements of the noise properties and a detailed comparison had been performed [89]. As reference cryo cooler a Leybold Vacuum (OLV) CP 5/100 were used.

In summary, the best cooler was selected that offers substantial advantages in flexibility and with the best compromise for a low vibration device combined with high cooling power. To illustrate the result, figure 3.5 presents the noise spectrum of the CP 5/100 compared with noise spectrum of the chosen cooler, OLV COOLPOWER 10MD (1 Hz rotation speed) as measured by G. Jacob [89]. The figure presents the amplitude of the noise (in acceleration of gravity) versus the frequency (in Hz). In contrast to the CP 5/100, the 10MD cooler has a pure sine spectrum with a moderate peak at the rotation frequency and low amplitudes at the side band frequencies. It becomes clear, that the 10MD has a noise floor that is more quiet than the other cold

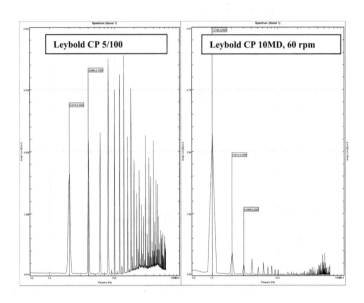

Figure 3.5: Measured noise spectra of two different cryo coolers. The Leybold CP 5/100
is the default cooler of the installed cryo pump. The Leybold CP 10MD is a
cooler that were especially developed for noise sensitive scientific instruments.
Therefore the 10MD cooler is used as cooling head in the used cryo pump. The
measurement has been performed by European Southern Observatory (ESO)
[89].

head. Due to the sinusoidal shape of the cooler noise, the further noise shielding
can be optimised on this specific frequency. Consequently, the effort that have to be
used for the further noise shielding becomes small. Another advantage of the 10MD
is that the rotation speed of the cold head is changeable. It can be varied between
0.33 Hz to 2.66 Hz. The cooling power and the noise generation are dependent on
the rotation frequency. At higher frequencies the cooling power is higher, it offers
almost twice as much cooling power as the CP 5/100. Whereas at lower frequencies
the noise level becomes minimal [89].

Additionally to the cryo cooler characterisation, ESO developed different techniques to reduced the vibrations in cryo cooled scientific instruments, like active
noise cancelling or passive damping [90]. Based on the published ESO results and
the Airbus heritage in highly precise metrology, a low noise cryo pump had been

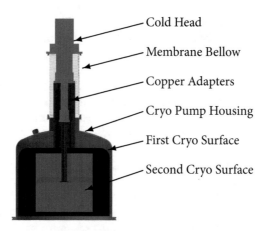

Cold Head

Membrane Bellow

Copper Adapters

Cryo Pump Housing

First Cryo Surface

Second Cryo Surface

Figure 3.6: Cross sectional snap shot of the custom cryo pump system. The CL 10000 cryo stages has been combined with a COOLPOWER 10MD cryo head which is additionally isolated via a 50 stage membrane bellow from the vacuum tank structure.

developed for the Airbus test facility, based on a COOLVAC (CV) 10000 CL cryo pump from OLV which had been modified. The standard CP 5/100 cold head had been replaced by the OLV COOLPOWER 10MD. Additionally, a 50 stage membrane bellow was used as connection between the vacuum tank and the cooler to decoupled the pump from the vacuum tank.

A cross sectional snapshot of the pump assembly is presented in figure 3.6. The pump consists of the COTS CV 10000 cryo surfaces, the CV 10000 housing, a 10MD cold head, a 50 stage membrane bellow and copper adapters. The cold head is shown in blue. The first stage is presented as black surface. During operation the first stage is temperature stabilised to 65 K. A Pt100 element is used to measure the temperature of the stage. The second stage is illustrated as green surface. The temperature of the stage is usually around 6 K. A Si-diode is used to determine the temperature of the stage. Both stages are able to pump various gases, e.g. the pumping speed for H_2O is $3 \cdot 10^4$ l/s, for nitrogen it is 10^4 l/s and for xenon about 8000 l/s. Heaters are mounted on each stage to enable a fast and controlled regeneration of the stages.

The copper adapters are used to extend the flanges of the cold head. They are presented in brown. Indium spacers are used between the single flanges to improve the heat transfer between the adapters. Thus, an optimal heat flow is guaranteed.

The membrane bellow that is used to decouple the pumping head is shown in yellow. To avoid a contraction of the membrane bellow because of the vacuum pressure, the cold head is mounted on an aluminium structure. The structure allows an adaptation of the initial load of the membrane bellow, i.e. the transfer function of the bellow can be optimised in reference of the rotation frequency of the cold head.

The whole cryo pump setup is controlled by a dedicated controller. The controller is connected with temperature sensors and the heaters that are part of the cryo stage. In reference to the measured temperature and the operation state of the pump the controller can turn the heaters on and off separately. The controller is also measuring the pressure inside the pump with a specific manometer. Moreover, the controller can be commanded via RS-232 from another device.

At the beginning of this chapter, in figure 3.2 the controlling of the installed pumps were illustrated. All pumps and the manometer can be controlled via the workstation. The turbo pumps are connected via a RS-484 bus to the workstation. All functions of the turbo pumps can be managed with the workstation. The manometer is also connected to the workstation, thus the vacuum pressure is always locked and stored on the workstation, i.e. the background vacuum pressure value is part of all performed measurements. The pressure readout is also used to control the vacuum pumps. The cryo pump is also connected to the workstation.

Due to the connection of all pumps to the workstation an autonomous evacuation, regeneration and venting of the vacuum vessel can be performed.

3.2 Plasma Diagnostic Setup

As previously presented in figure 3.1, the plasma diagnostic assembly consists of a set of 15 Faraday cups and one RPA. The whole plasma diagnostic setup was developed to be able to perform a thruster characterisation of the most important EP thruster parameters to complete the data which is generated by the micro-Newton thrust balance. Moreover, to be able to continue the work of A. Keller [32] where only plume diagnostics were used for thruster characterisation, a similar plasma diagnostics is required. Hence, the instruments of the plasma diagnostic should be comparable to the devices which were used in the mentioned study to maintain the continuation of data. Therefore, the developed probes are similar to the probes which had been developed at the University of Gießen by H.P. Harmann [40], W. Gaertner [91] and P. Koehler [34, 92].

The self developed instruments are placed on a jib arm. The arm can be rotated with a stepper motor from $+90°$ to $-90°$ around the thruster exit which shall be characterised. The jib arm is made of Item Profile 5 profiles. Therefore, the arm can simply be adapted on the specific needs of the actual measurement campaign, i.e.

the distance of the probes to the thruster can be varied or the lateral position of the probes can be adapted. For example, RITs have usually a collimated ion beam and hence a high divergence efficiency. Thus, the lateral spread of the probes can be reduced to increase the lateral resolution. Whereas, HEMPTs so far have a less efficient beam divergence, i.e. a wider spread of the probes is required in order to characterise the complete thruster plume.

The structure of the jib arm is also designed to be lightweight and maximum stiff. The light structure leads to a reduction of the momentum which has to be created to move the arm. The high stiffness of the arm is required to avoid oscillations of the arm which would decrease the angular measurement resolution.

The stepper motor (Phytron VSS 52) that is used to move the diagnostics around the thruster has a resolution of 200 steps per 360 °. Due to the digital controller of the stepper motor, a step can be digitally resolved as 1 : 256. The stepper motor is connected to the jib arm via a transmission. The transmission has a gear ratio of 1 : 256. Hence, the resolution of the jib arm is $2.7 \cdot 10^{-5}$ °. The whole stepper motor assembly was designed and manufactured by Phytron GmbH.

The rotation axis is always set inside the thruster exit plane (see figure 3.9). Because of the flexible Item structure the position of the rotation axis can be adapted to the specific needs of the actual test campaign. In general, it would be possible to rotate the arm more than $+90$ ° to -90 ° around the thruster, but due to the actual physical dimensions of the arm (725 mm long) and the dimension of the vacuum chamber it is currently not possible.

The output signals of each detector are transferred via coaxial cables to the outside of the vacuum tank where the analogue signals are digitalized. The whole data acquisition and probe controlling configuration is illustrated as a block diagram in figure 3.7. The ion current which is received by the Faraday cups or the RPA collector is directly transferred into a voltage signal behind the detector and amplified on a sufficient voltage level. The analogue signals are transferred to the outside of the vacuum chamber. The signals are low-pass-filtered (20 Hz, RC-filter) to remove high frequency noise and to avoid aliasing. Behind the filter the signals are digitalised. A National Instruments data acquisition board (NI PCI-6289) is used to digitalise the signals. The board provides 24 - 17-bit ADCs and 4 - 17-bit Digital Analogue Converters (DACs). The data acquisition board is also used to control the high voltage power supply that powers the retarding grid of the RPA. The maximum output voltage of the power supply is 6000 V. The output of the power supply is directly connected to the retarding grid of the RPA (yellow line in figure 3.7). The stepper motor is controlled by the workstation. The engine has a dedicated controller which is connected via USB to the workstation. The controller itself generates the required input signals for the stepper motor.

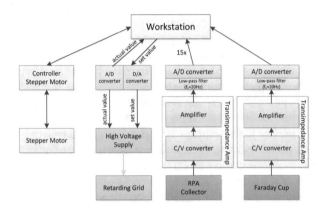

Figure 3.7: Flow chart of the plasma diagnostic setup. The whole setup is controlled via the workstation. The measured ion current is transferred into a voltage signal due to a transimpedance amplifier and converted into a digital signal with a set of Analogue Digital Converters (ADCs). The position of the jib arm is also digitally controlled by the workstation. Thus, fully autonomous measurement runs can be performed.

Due to the centralised control architecture of the plasma diagnostic, an autonomous execution of a full 180° plume characterisation can be performed. The acquired data is directly stored on the workstation. After the completion of the diagnostic run, the stored data is computed with Matlab. In the following subsections an introduction of the developed probes will be given.

3.2.1 Faraday Cup

On-board of the jib arm 15 Faraday cups are placed. The cups are separated into groups of three cups. Each group forms its own Faraday cup diagnostic assembly that can be operated independently from the other cups. Figure 3.8 presents one of these Faraday cup assemblies. The assembly consists of the three Faraday cups which have an inner diameter of 12 mm and a length of 30 mm. The bottom of each single cup has a tapered ending. It is surrounded by a plastic isolator made of POM (also called Delrin). The plastic is used to isolate the cups from each other and the ground. The POM is also used as mechanical interface to the jib arm. On top of the plastic isolator an electron repelling shield is placed. The POM structure isolates the shield from the cup and the ground. The electron repeller is made of aluminium. The diameter of the aperture of the electron repeller is 5 mm. Since the temperature of electrons inside the plasma plume is in the order of some eV [93, 2], it is sufficient to bias the repeller with -15 V in order to prevent the detector from electron influx. This is important to ensure that only the ion current will be detected. At the backside of the POM isolator the electronic circuit is placed. The Printed Circuit Board (PCB) is electrically contacted via a single M3 screw to the cup. Every cup has its own screw. The screw is also used as mechanical fixation of the cup to the isolator and the PCB.

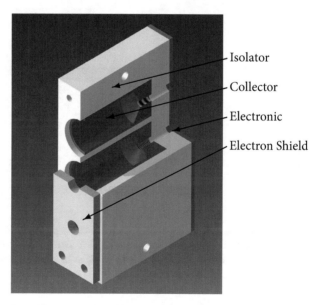

     Isolator
     Collector
     Electronic
     Electron Shield

Figure 3.8: CAD snap shot of the developed Faraday cup unit. Three Faraday cups are housed in a single unit. The cups are born by a POM housing. A negative charged shield in front the cups shields the collectors from electron injection.

On the PCB the same circuit is implemented three times, so that every cup has a specific transimpedance amplifier, filter and gain amplifier. In default configuration the cup is able to measure a current between $0\,\mu A$ and $2\,\mu A$. The current is transferred to a voltage signal in the range of $0\,V$ to $10\,V$. Thus, the theoretical resolution is $15.3\,pA$ with reference to the used Data Acquisition Board (DAQ) board. However, this resolution cannot be achieved because of various noise sources. This influences will be discussed in section 3.2.3. Like the whole micro thruster test facility, the circuit is designed to be adjustable, i.e. via the replacement of a single resistor on-board the PCB, the sensitivity of the probe can be adapted. Therefore, the probe can be used to measure different kinds of ion thrusters in the range of some micro-Newtons up to several thousands of micro-Newtons.

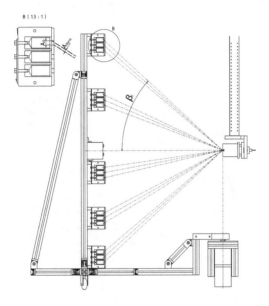

Figure 3.9: Drawing of the plasma diagnostic assembly to illustrate the geometrical distribution of the Faraday cups and the RPA. Due to the vertical orientation of the sensor a correction of the effective orifice area has to be performed with respect to the incidence angle β.

Figure 3.9 illustrates the diagnostic assembly as side view. The Faraday probe assemblies are attached vertically on the jib arm. In the middle of the drawing also the RPA shown. On the right side of the figure the thrust balance, the thruster under test and the stepper motor are illustrated. The cup assemblies are numbered from I to V. The first array (number I) is placed at the bottom end of the jib arm. The last array (number V) is placed at the top of the jib arm. The same numbering principle is used for the arrays itself, e.g. for the cup array I, the cup at the bottom is cup I.1 followed by cup I.2 and at the top of the array is cup I.3.

Since the orientation of the orifices planes of the cups with respect to the thruster exit is parallel, the current density measurement determines the data in cylindrical coordinates. Assuming that the thruster is a point source, the ions enter the cups with an incidence angle which is called β. Thus, the area of the detector aperture, also called projected area depends on the incidence angle. The projected area can be written as

$$A_{proj} = cos(\beta) \cdot A_{dec} \quad , \quad (3.1)$$

where A_{dec} is the area of the detector aperture which is 19.63 mm^2 for the presented cup design. For each cup a correction factor has been calculated which is part of the data evaluation tool for the plasma diagnostic. The tool is implemented in Matlab.

3.2.2 Retarding Potential Analyser

Figure 3.10: Picture of the developed RPA, without housing. The grids which are used to apply the retarding potential are mounted between the copper plates. The plates are separated by acrylic glass spacers.

To obtain information about the ion energy, a RPA is part of the plasma diagnostic setup. The device is usually placed in the middle of the jib arm as shown in figure 3.9. Because of the complexity of the device only a single RPA is part of the setup. Figure 3.10 presents a picture of the developed RPA. The analyser consists of three grids which are mounted and aligned via copper plates inside the detector. The grids are separated by acrylic glass spacers. Behind the grids a copper plate is used as ion collector. The collector is connected to a PCB where the measured ion current is transduced into a voltage signal. The circuit is almost identically to the circuit that is used for the Faraday cups. The housing of the detector is made of aluminium and protects the detector and the included electronics from sputter particles and electrical noise. The housing enables an adequate venting of the RPA. The single parts are locked with four Nylon screws. The screws and the acrylic glass spacer are also enabling an electrical isolation of the components.

In an ion energy detector the ion optics plays obviously a very important role. The size of the single grids and the distance between the grids have to be defined carefully in order to create a good separation of the ion energies, a maximum transparency of the grids, a good suppression of secondary electron emission and to avoid inadvertent physical effects. The most important theoretical aspects and dimensioning of the most parameters are presented in [91, 94]. Based on the published data, the specific design of the developed detector is summarised in [95].

Table 3.2: Summary of the geometrical parameters of the RPA.

Parameter	Value
Distance grid 1 - grid 2	7 mm
Distance grid 2 - grid 3	7 mm
Distance grid 3 - collector	1.5 mm
Diameter housing	15 mm
Diameter grid 1 & 2	12 mm
Diameter grid 2	12 mm
Wire diameter grid 1 & 3 and transparency	200 µm and 51 °
Wire diameter grid 2 and transparency	29 µm and 28 °

In figure 3.11 a sketch presents the parts of the RPA. Every single grid assembly consists of a stainless steel grid which is clamped between the two copper plates. In the middle of the copper plates is a hole that defines the aperture of the grid. The diameters of the holes of the copper plates are dependent on the specific grid. The dimensions of the grid assemblies are summarised in table 3.2.

As mentioned, the detector consists of three grid assemblies. The first grid assembly is shown in yellow in figure 3.11. It is called primary electron repeller, or grid 1. This grid has the same function as the aluminium plate of the Faraday cup, it should protect the detector from electron influx. The grid has typically a negative bias of -25 V. The grid is made of a 200 µm wire which forms a wire mesh. The distance of the wires inside the net are 700 µm. The net has a transparency of 51 %.

The second grid is called retarding grid or grid 2 (shown as red grid). The grid can be biased positively from 0 V to 6000 V. The grid is made of a thinner 29 µm

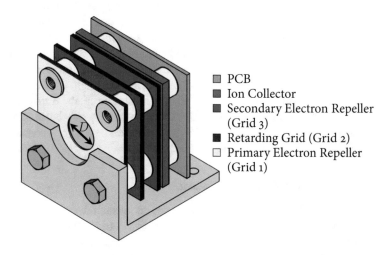

Figure 3.11: Sketch of the RPA to illustrate the parts of the assembly. The housing is sliced which allows a view of all RPA parts.

wire. The distance of the wires inside the wire mesh are 62 μm. The grid has a transparency of 28 %. This grid forms the RPA, since it slows down the incoming ions with respect to their energy. The RPA is designed that the incoming single charged ions cannot pass the retarding grid if the bias potential of the grid is equal or bigger than the ion temperature. For example, if grid 2 has a bias potential of about 300 V, only singly charged ions with a perpendicular velocity component corresponding to $\geq$ 300 eV would be able to pass the grid and is detected.

The blue grid assembly in figure 3.11 is called secondary electron repeller or grid 3. It is used to suppress the electrons which are potentially generated by the ions that are impacting on the ion collector. The dimensions of the grid are the same as the dimensions of the first grid. The grid is typically biased with −25 V.

All grids were selected in order to achieve an optimal and homogeneous electrical field inside the detector and a transparency which is as high as possible.

Behind the last grid, the copper made ion current collector is placed. It is shown in brown in figure 3.11. The plate collects the impacting ions and the current is transferred to the PCB board at the backside of the collector (presented in green).

Due to the grid system inside the RPA not all ions which enter the detector are reach the collector, i.e. the transparency of the detector is naturally smaller than 100 %. An estimation of the transparency has been performed to ensure that the real ion current density can be measured. This is of special importance for the presented setup since only the RPA is placed in the centre axis and thereby the RPA is used in parallel as ion current detector (see figure 3.9). To determine the transparency, different opportunities are thinkable like a theoretical estimation, an optical assessment, or the direct measurement in comparison to a Faraday cup. Considering that the uncertainties of a theoretical calculation would be high and that an optical measurement would requiring an extra setup, the transparency was estimated by a direct comparison with the developed Faraday cup that was temporary placed at the same vertical position. With respect to the specific apertures of the detectors the transparency was determined to 8.1 %.

3.2.3 Plasma Diagnostic Calibration

The experimental determination of the sensitivity of each detector is necessary to perform a precise measurement and as a first step to test if the detector works as expected. Hence, a calibration must be performed. Additionally to the calibration, an assessment of possible noise terms should be performed as well. To calibrate the detectors, a highly precise micro-Ampere source meter with a certified precision is used (Keithley 2401). The source has a resolution of 10 pA. To calibrate the detectors, the current source is directly connected to the current input of the probe, i.e. for the Faraday cups directly to the cup and for the RPA directly to the collector plate. In order to avoid leakage current to the ground, the sensing input of the source will also connected as close as possible to the detector ground. The result of a calibration run of one probe is illustrated in figure 3.12 where the applied current is plotted versus the measured voltage. The measurement points are shown as blue dots. To determine the sensitivity (in $V/\mu A$), a linear fitting was performed. The fit is shown as solid curve and the fitting parameters are presented at the left side. The curve demonstrates that the developed detector electronics is linear and that the detector works properly according to the design specifications. Like all measurement values, the acquired data of the plasma diagnostic setup has an uncertainty. Hence, an estimation about the expected uncertainties has to be performed to generate trustworthy measurement results. Different effects affect the accuracy such as the resolution of the data acquisition board, the noise of the electronics in general, the divergence of the ion beam, the influence of the plasma sheath on top of the detector surfaces, or impulse exchange of the particles inside the thruster. The grid system inside the RPA causes additional uncertainties like the mentioned transparency, lensing effects of the grid, potential deviation of the retarding grid or unknown secondary electron emission at the grid. A summary of known uncertainties are given in table 3.3. The influences of the other error sources are analytically hard to

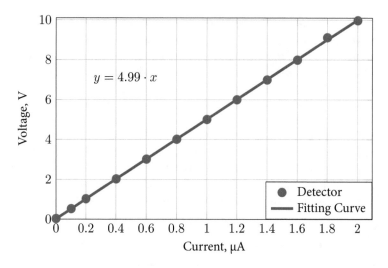

Figure 3.12: Result of the performed ion current detector calibration via a highly precise current source. The curve presents the measured voltage versus the applied current. The slope of the fitted curve is the sensitivity of the characterised device.

describe and to judge, therefore no specific value are given. It can be seen that only the measured electrical noise of the diagnostics limits the resolution to 600 pA.

Table 3.3: Summary of the determined detector uncertainties [95].

Uncertainty cause	Relative error
DAQ and electrical noise	0.05 %
Ion beam divergence	0.005 %
Potential deviation of retarding Grid	0.055 %
Lens effects	0.032 %

3.3 Airbus Micro-Newton Thrust Balance

The following section presents the Airbus Micro-Newton Thrust Balance as a major component of the hereby presented work. In the last years and in the framework of the presented work different thrust stand generations have been built. In the following, the micro-Newton Thrust Balance Mark III will be presented.

The balance is the fundamental measurement instrument of the micro-Newton thruster test facility. The thrust stand is especially designed to perform thrust noise measurements of highly precise AOCS thruster candidates for LISA or other future scientific space missions. Moreover, the balance is able to perform absolute direct thrust measurements. Currently the LISA mission concept has the most challenging micro-Newton thruster requirements in point of thrust noise and resolution. Thus, most balance requirements can be derived from the LISA AOCS thruster requirements (see table 1.1 [7, 9]). Additionally, the micro-Newton thruster requirements of NGGM had been considered due to the required throttability from $50\,\mu N$ to $2500\,\mu N$ [11]. The balance requirements are summarised in table 3.4.

Table 3.4: Summary of the thrust balance requirements which are derived from the LISA and the NGGM thrust noise and thrust range requirements

Parameter	Requirement
Thrust range	$0 - 2500\,\mu N$
Measurement bandwidth	$10\,Hz$ to $1 \cdot 10^{-4}\,Hz$
Thrust noise	$< 0.1\,\mu N/\sqrt{Hz} \cdot \sqrt{1 + (\frac{10\,mHz}{f})^4}$
Thrust resolution	$< 0.1\,\mu N$
Max. thruster weight	$6\,kg$

The design of the balance combines the Airbus heritage in highly precise laser metrology with a hanging pendulum balance. This idea was firstly demonstrated in the Diploma Thesis of F. Hey [48]. The performance of this thrust stand was not sufficient to perform thrust measurements in the sub micro-Newton regime and the requirements of LISA were not fulfilled, but a further development seemed promising.

To underline the principle, figure 3.13 presents the developed thrust balance. It consists of two hanging pendulums, a measurement pendulum and a reference pendulum. As translation and deflection readout a heterodyne laser interferometer is used. The laser beams between the interferometer and the pendulums are illustrated as thin (red) lines, the deflected pendulum is shown with dotted outlines. The thrust measurement is performed by a differential translation measurement of the pendulums, i.e. the translation of the measurement pendulum minus the translation of the reference pendulum multiplied with the specific sensitivity results in the measured thrust. Therefore, external noise is not taken into account.

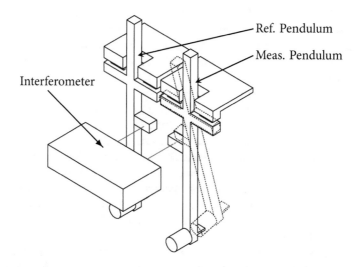

Interferometer

Ref. Pendulum

Meas. Pendulum

Figure 3.13: Sketch of the double pendulum balance principle. Two fully symmetric hanging pendulums are combined with a laser interferometer as translation sensor. The combination of a highly precise and highly stable optical readout with the double pendulum setup leads to a common mode rejection of various noise sources and hence to a highly stable thrust measurement device.

This principle is called common mode rejection and requires a symmetric assembly which enables that different parts act equally on external disturbances. Additionally to the symmetry of the mechanical parts of the system, all other components such as the optical setup and the electrical setup have to be symmetric as well. Thus, various noise sources are suppressed.

An examples for a possible noise source is seismic noise, due to the moving parts of the systems (pumps etc.), or staff that works around the laboratory. Seismic noise can directly lead to a translation of the pendulums that could be misinterpreted as a variation of the thrust. But, the disturbance will not be measured due to symmetry of the setup.

Another example is thermal noise because of temperature variation inside the room or symmetric temperature loads during thruster firing. This can cause drifting of the electronics used or variations in the optical path length of the interferometer or it can lead to mechanical deformation of the pendulums because of thermal expansion. However, assuming that both pendulum are affected in the same way, the differential measurement would not measure this noise.

It becomes clear that the unique idea to couple a highly symmetric laser interferometer with a total symmetrically mechanical structure is potentially able to suppress all sources mentioned, but also other unknown noise sources. Therefore, the present thrust balance setup is able to measure forces in the sub micro-Newton regime. Furthermore, the balance allows thrust noise characterisation within the whole LISA measurement bandwidth.

As previously pronounced, the main goal of the thrust balance developed is to enable a direct thrust noise characterisation of possible LISA AOCS thruster candidates. Also absolute thrust measurements should be performed. To fulfil both requirements, the developed thrust balance is able to operate in two measurement modes. They are called open loop and closed loop. In open loop the balance swings freely , thus the measured thrust corresponds directly to the pendulum deflection, or translation measured. In closed loop an actuator is used to hold the pendulum in place. Therefore, this mode can also be called in situ measurement mode. In this mode, the applied thrust is equal to the force of the actuator. Typically, the open loop measurement should be used for thrust noise characterisation, since no additional control loops or other electronics adds noise to the measurement data and the absolute resolution of the laser interferometer is negligible (see section 3.3.2). Whereas, for absolute thrust measurements a direct physical reference like the actuator force ensures correct absolute thrust values.

Additionally to the implementation of the required symmetry, the thrust stand should be able to measure different micro-Newton thruster candidates, independently from the principles on which the thrusters rely. Therefore, the thrust balance development was following a generic design approach. Most of the parts of the balance assembly can be rearranged and it should be simple to add new features to the balance.

In the following sections, a detailed overview of the mechanical, optical and electrical design of the balance will be given.

3.3.1 Mechanical Balance Setup

The requirements mentioned of the thrust balance lead to a specific mechanical design. Figure 3.14 provides an overview of the major parts of the balance. Some parts of the assembly are cut to improve the visibility of all parts of the assembly. The illustration presents the balance structure (in blue) that carries the pendulums (in green). On every pendulum a mirror is mounted (in red). The mirrors and the interferometer head (in grey) form the translation readout. The thruster and other additional hardware (in yellow) are mounted on the pendulums as well. The measurement pendulum bears the thruster assembly which shall be tested. Whereas, the reference pendulum bears usually mass dummies. The thruster setup which is

presented in figure 3.14 is the Engineering Model (EM) of the RITμX. It consists
of a RFG and the thruster. Due to the high thermal load of the RFG, a radiator
is mounted on top of the RFG (in brown). To suppress the first frequency of the
pendulum an eddy current brake assembly (in orange) is part of the thrust balance.
As actuator a voice coil (in dark red) is mounted on every pendulum. Counterweights
(bronze-coloured) are used to adapt the spring rate of the pendulums.

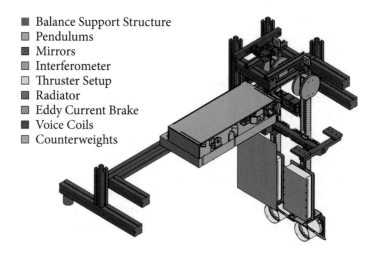

■ Balance Support Structure
▢ Pendulums
■ Mirrors
▢ Interferometer
▢ Thruster Setup
■ Radiator
▢ Eddy Current Brake
■ Voice Coils
▢ Counterweights

Figure 3.14: Illustration of the complete pendulum setup. The support structure (in blue),
the interferometer (in grey) and other parts are partly sliced to the view on the
other balance components, such as the pendulums (in green), the mirrors (in
red) and the eddy current brake (in orange).

It is essential that the structure does not influence the pendulum performance
and the performance of the electric thruster under test. Moreover, the structure
should be simple to adapted and to modify. Therefore the structure should fulfil the
following requirements:

- Highly stiff and highly stable

- High eigenfrequency which is not part of the measurement band

- Non magnetic

- Electrically conducting to avoid unknown space charges

- Simple to adapt

- Good handling possibilities

The structure of the balance is split into two parts, the minor support structure and the major support structure. The minor support structure is an aluminium framework made of Item profiles. Thus, an easy adaptation of the structure is feasible. Moreover, the framework is lightweight, stiff and stable.

The major balance structure carries the two pendulums and the whole pendulum bearing assembly. Figure 3.15 give an overview of the pendulum bearing assembly. The pendulums (in green) are connected via a set of leaf springs to the support structure. As long as the leaf springs are used in the elastic regime, the connection works as a quasi frictionless hinge. Moreover, the leaf springs can be used as data connection to the pendulums. The springs are mounted to the structure and to the pendulums via clamping. On one site a ceramic plate (in white) ensures that the springs are mounted in a defined plane to avoid constraining forces in other degrees of freedoms as expected. At the opposite site plastic plungers (in yellow) push the springs onto the ceramic plates. To use the springs as data and power connections, they have to be connected to the data and power harness of the facility and the thruster. For this purpose custom-built cable shoes (in orange) are used which are pushing the cables onto the leaf springs.

The total number of leaf springs can be adapted on the specific needs of actual measurement campaign. The minimal number of springs that have to be used is four leaf springs per pendulum. Whereas, the maximum number of springs is 20 leaf springs per pendulum, this configuration is presented in figure 3.15. The spring constant of the whole bearing assembly remains constant for all possible spring leaf configuration, because the overall surface area of the springs is kept constant.

For instance, if the thruster that shall be operated a high demand of electrical power is required, a lower but wider number of springs should be used in order to reduce the thermal load of a single spring.In the case that a test requires additional measurement equipment on the balance more springs can be used as data transmission lines, etc. Commonly, the bearing consists of 12 leaf springs. Two wide springs as thruster power feedthroughs and 10 thinner springs for data transmission, controlling and grounding.

Some electric thruster require a propellant feed line to the thruster since the tank and valves cannot be placed always onto the pendulum structure. Thus, a propellant tube has to be connected to the balance as well. As propellant feed line from the support structure to the pendulum a PTFE-tube is used. The tube is not shown in figure 3.15, but it is mounted on the backside of the ceramic plates, as close as possible to the pivot to keep the influence on the spring rate minimal.

As previously referred in section 2.4.1, the spring rate of a single pendulum consists of the spring rate of the bearing (K_s) and the spring rate of the hanging pendulum

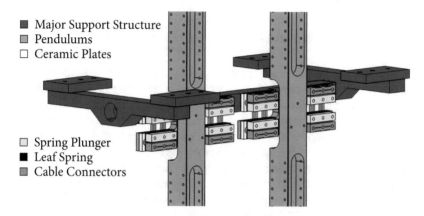

■ Major Support Structure
▨ Pendulums
□ Ceramic Plates

□ Spring Plunger
■ Leaf Spring
▨ Cable Connectors

Figure 3.15: Illustration of the pendulum bearing assembly. The pendulums are born via a set of leaf springs, the absolute number of springs can be varied between 4 and 20 springs per pendulum with respect on the thruster that shall be tested. In every case the stiffness of the bearing is kept constant. The leaf springs are used in parallel as data and power connections.

itself, caused by gravity. The springs of the bearing are connected parallel since they are arranged in a single plain. The length (l_{spring}) and the thickness (t_{spring}) of the leaf springs are equal. As mentioned, per definition only the width of the single springs and the number of springs can be varied. However, the sum of all leaf spring widths (w_{total}) is always defined as 20 mm. Thus, the spring rate of the whole bearing is not dependent on the used leaf spring configuration and the springs can be assumed as a single leaf spring. According to the bending beam theory [96], it can be calculate by

$$K_s = \frac{E \cdot w_{total} \cdot t_{spring}^3}{6 \cdot l_{spring}^2 \cdot \Delta l_{thruster}} \quad , \tag{3.2}$$

where E is the modulus of elasticity and $l_{thruster}$ is the distance between thruster and the centre of the pendulum bearing. In case that the CoG is not inside the centre of the bearing, which is the rotation point, the spring rate of the hanging pendulum caused by gravity has to be considered. Assuming that the pendulum has a smooth mass distribution it can be modelled as physical pendulum. Therefore, the spring constant of a single pendulum is defined by:

$$K_{pendulum} = K_s + \frac{m_p^2 \cdot g_0 \cdot l_{cog}}{I_{pendulum}} \quad , \tag{3.3}$$

whereas m_p is the mass of a single pendulum, l_{cog} is the distance between the CoG and the bearing centre and $I_{pendulum}$ is the moment of inertia of the pendulum.

The equation underlines that it is possible to vary the spring rate of the balance by changing the CoG of the pendulums. To adapt the spring rate, counter weights are used. Figure 3.16 provides an overview of the counterweights positions (presented in purple). Usually, the weights are placed at the ends of the pendulum arm to generate a maximal effect with a minimum counterweight mass. Therefore, coupling between the change of the CoG and the variation of the moment of inertia is minimised. Moreover, the low counterweight mass leads to a small total mass of the pendulum itself which results in a higher balance sensitivity at higher frequencies because of the higher pendulum eigenfrequency (see section 2.4.1).

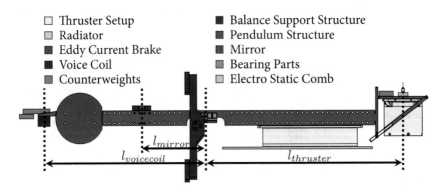

Figure 3.16: Illustration of a single pendulum sub assembly which is rotated by 90 degree. On the pendulum structure (in green) all other required balance parts and the thruster that shall be tested are mounted. The figure illustrates the distances between the pendulum pivot and the thruster, the mirror and the voice coil.

Figure 3.16 provides a full overview of a single pendulum. The illustration is rotated by 90 ° and allows a detailed look on all parts of the pendulum assembly. In dark green the pendulum structure is shown. The balance support structure is drawn in dark blue. The thruster and other thruster components (RITµX together with the RFG) are presented in yellow. To increase the thermal stability of the pendulum and to cool the RFG, a radiator is mounted on the RFG (in bright blue). The conducting plates of the eddy current brake is shown in orange. To actuate the pendulum (see section 3.3.3 for more details) a voice coil is mounted at the end of the pendulum structure. The counterweights which are used to adapt the CoG are drawn in pink. The mirror that reflects the interferometer light is presented in red. The bearing assembly is shown in grey. To calibrate the thrust balance an ESC is mounted on each pendulum, more details about the calibration will be given in section 3.3.4. The figure also illustrates the distances between the pivot and the thruster ($l_{thruster}$), the mirror (l_{mirror}) and the voice coil ($l_{voicecoil}$). The specific values for the presented setup are provided in table 3.5.

Table 3.5: Summary of a representative case of the physical parameters of one single pendulum with the RITμX mounted on the pendulum. Parameters like $l_{thruster}$, or the thruster weight are dependent on the thruster that shall be tested. Thus for each possible configuration a new assessment of the pendulum parameters have to be performed.

Parameter	Specific Value
Distance pivot to thruster ($l_{thruster}$)	397 mm
Distance pivot to mirror (l_{mirror})	125 mm
Distance pivot to voice coil (l_{mirror})	320 mm
True leaf spring width (w_{total})	20 mm
Leaf spring thickness (t_{spring})	0.15 mm
True leaf spring grip of rivet (l_{spring})	9 mm
Pendulum assembly weight(excl. thruster)	3.4 kg
Thruster assembly weight	1.4 kg
Position CoG from bearing axis (l_{CoG})	1 mm
Moment of inertia ($I_{pendulum}$) by l_{CoG}	423329 kg mm^2

Based on the data which were given in 3.5 and the equations 3.2 and 3.3, an estimation of the balance sensitivity can be performed. The sensitivity is given by the ratio of l_{mirror} and $l_{thruster}$ and the spring rate. It can be written as

$$S = \frac{l_{mirror}}{l_{thruster}} \cdot K_{pendulum} \qquad . \tag{3.4}$$

The calculated spring rate of the assembly is 0.0705 μN/nm. The ratio between l_{mirror} and $l_{thruster}$ is 0.315 hence, the calculated sensitivity is 0.022 μN / nm. Due to the modular design of the pendulum, the given values can be easily adapted to the requirements of the actual measurement campaign, e.g. if a higher sensitivity is needed the CoG can be changed to fulfil those requirements. Also a negative CoG is applicable as long as the system is in a stable state.

Since the spring rate of the pendulum is known, an estimation of the eigenfrequency can also be performed in respect of the equations 2.20 and 2.23. The computed eigenfrequency is 0.811 Hz. It is inside the targeted measurement bandwidth, which is typical for a pendulum with a mass of some kilograms. However, due to the implemented eddy current brake the low eigenfrequency will not compromise or limit the measurement results. The results are summarised in table 3.6.

Table 3.6: Summary of the estimated pendulum spring rate, eigenfrequency and the derived open loop sensitivity.

Parameter	Specific Value
Estimated spring rate	0.0705 µN/nm
Estimated pendulum eigenfrequency	0.811 Hz
Derived balance sensitivity	0.022 µN/nm

3.3.2 Optical Balance Readout

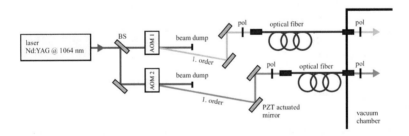

Figure 3.17: Sketch of the laser setup that is used to generate the heterodyne frequency, called frequency generation [8].

As presented in section 2.4.1, the readout of the thrust balance has a serious impact on the overall thrust stand performance. With reference to the given requirements of the balance and the Airbus heritage in highly precise metrology [50, 53, 8, 49], a heterodyne laser interferometer according to Wu et.al. [97] is used as balance translation and deflection readout. In opposite to a classic laser interferometer two laser beams with a defined frequency difference, so called heterodyne frequency are superimposed in front of a photodiode. The measured signal of the photodiode is the sinusoidal heterodyne frequency. The variation of the phases of the acquired signals are corresponding with the relative translation of the laser beams.

The laser interferometer system consists of the laser interferometer head, the frequency generation, the laser source and all required electronics e.g. photodiodes, amplifiers, filters and the digital phasemeter. Figure 3.17 provides an overview of the optical system parts which are placed outside the vacuum chamber, they are called the frequency generation. As laser source a Coherent Mephisto S Nd:YAG with non-planar ring oscillator and 200 mW output power at 1064 nm is used. Behind the laser the beam is split by a 50:50 Beam Splitter (BS). To generate the heterodyne frequency, an Acousto Optic Modulator (AOM) per beam is used. The frequency of the first beam is shifted by 80.00 MHz (called f1) whereas the second beam is shifted by 80.01 MHz (called f2) therefore, the heterodyne frequency used is 10 kHz. The frequencies are generated by a Direct Digital Synthesiser (DDS). The frequency generator is locked to the global reference clock.

After the modulation of the beam frequencies the beams are coupled into optical fibres to transfer the light into the vacuum chamber and to the laser interferometer. In order to avoid errors caused by polarisation changes polarisers are placed in front of the fibre collimators. One of the mirrors inside the optical path in front of a fibre coupler is actuated in one degree of freedom by a piezo electrical actuator. The mirror is used to compensate the variation of the optical path length inside the optical fibre due to temperature variations. This compensation is implemented as a digital control loop, called phase lock loop.

The interferometer is placed (see figure 3.14, 3.13) in front of the pendulums. An image of the interferometer with overlaying optical path is presented in figure 3.18. The beams which are launched from the fibre couplers (not shown in figure 3.18) are split by Koesters prisms (marked as K-Prism) to generate the reference and the measurement beam (marked as Ref. B. and Meas. B.). The f1 beams are reflected by a Polarising Beam Splitter (PBS) because they are s-polarised. After passing a $\lambda/4$ plate the beams are leaving the interferometer to the mirrors which are mounted on the pendulum arms (presented in figure 3.13). The beams are running perpendicular to this mirror i.e. they are reflected in itself and crossing the $\lambda/4$ plate one more time hence, the beams are p-polarised. Thus, the measurement and the reference beam are transmitted by the PBS and split by a 50:50 BS.

The f2 beams are guided via two mirrors directly to the 50:50 BS. A $\lambda/2$ plate is used to change the polarisation of the beams to s-polarisation. The beams are superimposed behind the BS in order to create the heterodyne frequency. One QPD per beam is used to transfer the interfered laser light into an electrical signal of the heterodyne frequency. The beams which are not used are guided into a beam dump in order to avoid optical cross talk.

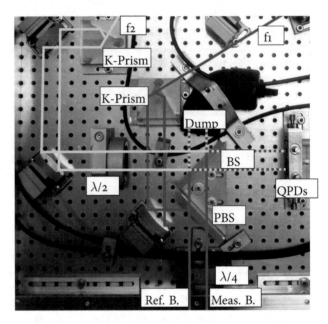

Figure 3.18: Picture of the laser interferometer with overlaying sketch of the laser beam
path. The laser beam f1 is superimposed with the laser beam f2 in front of the
Quadrant Photodiode (QPD) to generate the heterodyne signals of reference
and the measurement beams. The phase difference of the beams is equivalent to
the differential translation of the pendulums. Due to the QPD also the tilt of the
wave front of the laser beams can be determined, therefore an independent tilt
measurement of each pendulum can be performed.

The two optical heterodyne frequencies are transferred into an electrical signal by
two QPDs. Behind the QPDs a set of amplifiers is used to process the signals. The
signal flow is illustrated in figure 3.19. Inside the workstation a Field Programmable
Gate Array (FPGA) board is embedded that is used to process the acquired data
in real time. The FPGA is driven by an external 40 MHz reference clock. Inside
the FPGA the digital phasemeter is implemented. The phasemeter extracts the
phase information of the acquired frequencies via mixing with a 10 kHz reference
frequency. The phasemeter is executed with 200 kHz. Inside the FPGA a Look
Up Table (LUT) generates the reference frequency that is used as input for the
phasemeter as well as input for the phase lock loop, sometimes called phase controller.
Due to limited resources on board of the FPGA, the further transformation of the

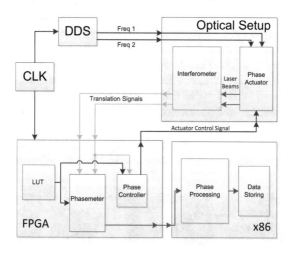

Figure 3.19: Flow chart of the interferometer setup. A DDS is used to generate the frequency offset between the laser beams. The acquired translation signals are processed in a digital phase meter. In order to avoid non linearities and to compensate the path length variations of the optical fibres, that are used to feed the light into the vacuum chamber, a phase controller is implemented.

phase signals is performed directly onto the workstation. The data is transferred from the FPGA to the workstation by a First In First Out Memory (FIFO). To reduce the quantity of data that must be transferred through the FIFO, the data is downsampled in front of the FIFO via a 120 Hz digital third order Butterworth filter. The data transmission through the FIFO is clocked with 400 Hz.

Figure 3.20 illustrates the electrical and digital signal flow of one of the eight acquired signals. In a first step the photo current is converted into a voltage signal by a transimpedance amplifier. Behind this amplifier the signal is filtered with a RC-filter and transferred to a symmetric signal. The amplifier stages and the filters are directly placed behind the QPDs, in the inside of the vacuum chamber. This part of the electronic is called front end. The signals are fed to the outside of the vacuum chamber via symmetric data transmission.

Before signals are digitalised by an ADC, another amplifier is adjusting the voltage levels. The signals are filtered before they are digitalised in order to avoid aliasing. The ADCs have a symmetric input. The symmetric signal transmission is enabling a low noise transferring of the acquired analogue signals which is required to fulfil the targeted noise requirements. The described stage is called the backend.

The digitalised signals are transferred via a serial peripheral interface to the workstation. The serial peripheral interface is clocked with 4 MHz. The ADC runs at a sampling frequency of 200 kHz. Inside the FPGA the signals are mixed with the reference signal sine and cosine. After the mixing of the signals, they are filtered with the mentioned 120 Hz digital third order Butterworth filter to eliminate higher orders and to reduce the information inside the signals to a sufficient level. The filtered signals are transferred via a FIFO to the workstation, where an arctangent processes the information to the phase of the acquired signals. This computed phases can be transferred into the translation output which is usually used.

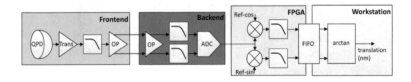

Figure 3.20: Diagram of the electrical signal flow. The photo current is transferred into a voltage signal, low pass filtered and transduced into a symmetric directly behind the QPD. This setup is called frontend and it is placed directly inside the vacuum tank. The backend translates the analogue signal into a digital signal. The digitalised signal is processed on the FPGA and on the workstation.

In general, the interferometer offers two measurement modi. Firstly, the measurement of the balance translation and secondly the measurement of the pendulums angles. The differential of the phases of the two laser beams is equal to the relative movement of the measurement and the reference mirror which are mounted onto the pendulums. Thus, a differential translation measurement of the two pendulums can be performed. This measurement mode is called translation measurement.

As mentioned, it is also possible to measure the deflection of the measurement and the reference mirror. This measurement mode is called DWS measurement. The tilt of the wavefront of the reflected beam in reference to the sensor plane can be measured by comparing the phase at two positions over the beam cross section [8, 98]. The used QPDs provide four measurement points, therefore the angle of the pendulum can be measured in two directions. The principle is explained in figure 3.21. On the left side of the sketch the case with not tilted measurement mirror is shown. The wavefront has no angle in reference to the sensor. Whereas, on the right side the case with tilted measurement mirror is presented.

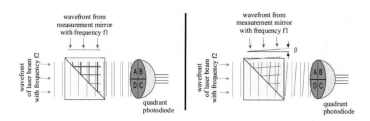

Figure 3.21: Illustration of the DWS. On the left side the not tilted case is shown in comparison with the tilted case that is presented at the right side [8].

It must be underlined that the translation measurement cannot be performed independently for the two pendulums because it is a differential measurement. Moreover, the absolute measurement range of interferometer is equal to the halved wavelength i.e. 532 nm. To increase the relative measurement range, an unwrap function is implemented by counting phase jumps.

In contrast to the translation measurement, the DWS allows an absolute and independent measurement of the pendulum angles up to several micro radiant. For both modi, the maximum measurement range is limited by the quality of the superimposition which can be described by the contrast. As result of the pendulum deflection also the mirror on the pendulum is deflected and therefore the beam is not back reflected perfectly in itself any more. Dependent on the path length between mirror and the position of the superimposition the contrast changes. Typically, the translation measurement can also be performed at low contrast levels of only some percent. But, the DWS is non linear if the contrast is smaller than fifty percent.

As previously explained, the optical readout is based on the Laboratory for Enabling Technology (LET) interferometer technology. A resolution of 5 pm in a bandwidth from 1 Hz to 10^{-2} Hz has been demonstrated [8]. Therefore, the theoretical thrust balance resolution can be assumed as 0.1 nN in the bandwidth mentioned and with a targeted spring rate of 0.02 µN /nm. Thus, the theoretical resolution is almost 1000 times smaller than required. However, due to the various noise source and the pendulum design itself the real resolution will be some orders of the magnitudes higher.

3.3.3 Thrust Balance Controlling and Data Acquisition

The major components of the thrust balance, their electrical connections and the grounding are presented in figure 3.22. As already explained, a workstation is used as central processing and controlling unit. The workstation is frequency locked to a

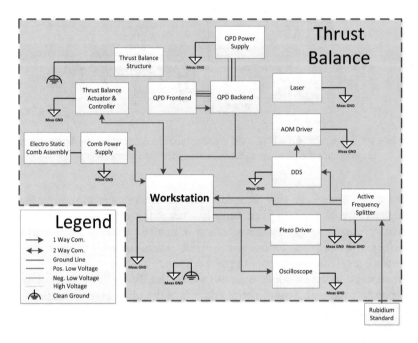

Figure 3.22: Block diagram of the micro-Newton thrust balance assembly in order to illustrate the major components of the thrust balance. All sub-assemblies are controlled by the workstation. The workstation includes also a FPGA that is used for the digital phasemeter and for the digital controllers that are required to operate the balance.

global reference clock. The reference clock is generated by a Rubidium Standard. To improve the long term stability of the rubidium standard, it can be locked to the Global Positioning System (GPS) timing reference. The DDS that drives the AOMs which are used to generate the heterodyne frequency is also connected to the reference clock.

The optical readout is connected via the frontend and the backend to the workstation. The acquired data is used as input for the phase lock control loops which is required for the laser interferometer. The controller is implemented as part of the phasemeter inside the FPGA. Via a DAC the set value is transferred to an analogue signal that controls the piezo driver.

As mentioned at the beginning of this chapter, the thrust balance is able to operate in open loop and closed loop. In open loop, the acquired data from the interferometer can be directly converted into the measured thrust. But, due to the limited absolute resolution of the interferometer a consequent nulling of the balance is required and in the case of fast variations (faster than the Nyquist frequency) of the measured thrust errors can occur.

In closed loop, the acquired interferometer information is used as input for the controllers of the pendulum actuators. To be precise, the DWS signal is used because it is independent for each pendulum. On each pendulum a voice coil is used as actuator. The voice coils enable a linear and effective actuator. For each voice coil a Keithley micro-Ampere source is used as voice coil driver. The current sources are controlled via IEEE-488.

The angle of the mirrors that or mounted on the pendulums can be actively steered in two degrees of freedom to realign the quality of the interferometer superimposition measured by the contrast. To steer the mirrors, Attocube piezoelectric driven positioners are used. The actuators are controlled by a specific piezo controller (includes the piezo drivers). The controller is connected with 4 feed lines to the mirror actuators, i.e. 4 leaf springs per pendulum are typically reserved for this connection. The workstation is connected with the piezo controller via USB. Hence, an autonomous realignment of the mirrors can be performed.

Also part of the electrical balance setup is the calibration device electronics. A detailed overview of the calibration setup will be provided in section 3.3.4. The high voltage power supply which is used to drive the calibration device has an analogue input that is used as interface between the workstation and the calibration device. Thereby, a fully automated calibration of the pendulum assembly can be performed.

An oscilloscope is also part of the setup as additional output to the qualitatively validation of the acquired frequencies and other data.

On the workstation all functions of the thrust facility are controlled by a LabView programme which contains all required sub-programmes and builds the interface to possible additional hardware. LabView is a graphical system design tool and development environment from National Instruments. The programme handles all data available, including the storing and commissioning of the data. The data can be plotted directly onto different screens to control the facility as well as the operated thruster. The acquired data is written into different text files. The files are synchronised by a unique time stamp, that is generated inside of the FPGA.

For further data processing a Matlab tool chain was developed. In a first step the stored data is translated from the text files to the Matlab workspace. After this step the processing of the data is performed.

The thrust measurement itself is defined as a standard procedure. Figure 3.23 provides an overview of the procedure. As first step of the measurement a commissioning is performed to ensure that the measurement conditions are sufficient and all instruments are working as expected. As second step a calibration of the thrust balance has to be performed. Firstly, the calibration is used to measure the real sensitivity of the balance. Secondly, the calibration allows to cross check if the thrust balance is as expected e.g. is the accuracy, the precision, or the repeatability is okay, etc.

The calibration is usually performed by the integrated ESC assembly (see section 3.3.4). But, dependent on the thruster type also the cold gas thrust of the thruster can be used for the calibration. In general, also the voice coils could be used for the balance calibration. As third step, the thrust measurements can be performed as planned. After the thrust measurement another calibration run is always performed in order to gain information about potential drifts of the thrust stand.

Figure 3.23: Flow chart of the standardised measurement run. Before and after the real thrust measurement a calibration is performed.

The Matlab data processing follows the same process. As first step the acquired data will be checked manually. The criteria are if the data has the right format, if the signal to noise ratios are as expected which also includes the amplitude of the signals and if the contrasts of the interferometer signal is sufficiently. After this the data from the calibration is analysed and the sensitivity is determined. Typically, the sensitivity is estimated via plotting the applied force against the measured translation. The slope of the plotted linear curve is equal to the sensitivity. Usually, more than one calibration is performed to avoid random errors. With respect of the obtained sensitivity, in open loop, the measured translation can be directly transformed into the measured thrust according to

$$F_{meas} = S_{calOL} \cdot \Delta x \qquad . \qquad (3.5)$$

If the measurement was performed in open loop a nulling of the data has to be performed during the data processing to determine absolute thrust values because of the limited absolute resolution of the interferometer. In closed loop a nulling is not required.

For closed loop the sensitivity is not given in µN/nm, since the output measurement signals are the input signal for the voice coils, i.e. in closed loop the sensitivity is given in N/µA. The measured force can be calculate by

$$F_{meas} = S_{cal_{vc1}} \cdot I_{vc1} - S_{cal_{vc2}} \cdot I_{vc2} \qquad , \qquad (3.6)$$

where S_{calCL1} is the sensitivity of the measurement pendulum, S_{calCL2} is the sensitivity of the reference pendulum, I_{vc1} is the input current of voice coil 1 and I_{vc2} is the input current of voice coil 2.

3.3.4 Balance Calibration and Uncertainty Analysis

In chapter 2 the fundamentals of thrust measurements by a pendulum balance has been explained. It became clear that the balance spring rate, or the sensitivity must be known in order to determine the measured thrust (with respect to equation 2.20). In section 3.3.1 a theoretical assessment of the balance spring rate was presented. However, an exclusively theoretical consideration of the spring rate produces large uncertainties in the measurement result, because of manufacturing and assembling tolerances, etc. In practice, the spring rate is determined by a calibration routine. A simple and most common way to determine the sensitivity of the thrust stand is an end to end calibration, i.e. a test from the mechanical thrust stand to all other devices until the data recorder. In parallel to the sensitivity quantification, the calibration can also be used to test the accuracy, precision, linearity and repeatability of the thrust balance. Moreover, the calibration can be used to measure the thrust balance response.

In [73] different thrust stand calibration techniques are discussed. In summary, the calibration can be performed by a known force, or by a known impulse. But, the know impulse method is only applicable if the balance is able to oscillate at their first eigenfrequency. Whereas, the first eigenfrequency shall totally be damped in the developed thrust stand, i.e. only the known force calibration can be performed.

Conventionally, hanging weights can be used to calibrate the thrust stand which are connected to the pendulum [43, 44, 99]. Those devices are best practise for calibrations in the milli-Newton range. But, small errors in the micro-Newton regime, caused by friction in the hanging weight bearing, are potentially falsifying the calibration. Therefore, calibrations where the calibration device requires physical contact to the pendulum are not sufficient accurate and repeatable enough to calibrate force in the sub-micro-Newton regime.

In order to overcome these limitations, different contactless methods have been developed and introduced. Examples of those devices are capacitors in different variations, voice coils or gas dynamic calibration techniques. Simple plate capacitors are usually not used because the applied force depends strongly on the plate distance. Voice coils require an additional calibration and characterisation of the calibration device itself. And gas dynamic calibration requires numerical gas flow simulations to estimate the force produced or a calibration of the calibration device with another thrust balance.

In [79] an experimental comparison of two different calibration devices had been performed. The novel micro-mechanical comb actuator, called ESC which was introduced by Johnsan and Warne [100], was compared with a precise gas dynamic calibration technique that relies on the free molecule expansion of a gas through a thin-walled sonic orifice [101]. In summary, it has been demonstrated that both techniques are able to calibrate thrust stands in the sub-micro-Newton regime. The differences between the techniques was not more than 8 % within the whole measurement spectrum (100 nN to 1200 nN).

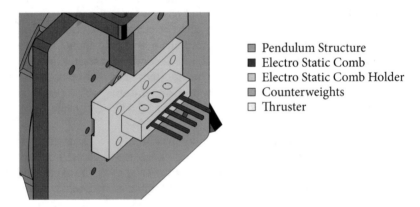

- ▨ Pendulum Structure
- ▮ Electro Static Comb
- ▢ Electro Static Comb Holder
- ▨ Counterweights
- ▢ Thruster

Figure 3.24: Illustration of the one side of the ESC assembly. It is directly mounted at the backside of the thruster, thus no conversion between the calibration force and the measured thrust of the thruster has to be performed.

With respect to the unique advantages of the ESC such as simplicity, wide operation range, insensitivity to balance motion, the ESC was selected as standard calibration device. Figure 3.24 presents the pendulum side of the comb assembly which is part of pendulum assembly. Each pendulum bears an own ESC. The assemblies are placed directly at the backside of the thruster, inside the symmetry axis (see figure 3.16). Therefore, no further correction because of the different leather arms has to be performed after a calibration cycle.

The full calibration assembly consists of two combs. Each comb is clamped into a comb holder. One comb holder is mounted on the pendulum. Whereas, the other comb holder is mounted on a xyz-translation stage that is mounted to the pendulum support structure. The comb on the pendulum remains always fixed. The comb on the xyz-stage is used to align the whole assembly. A schematic of the comb assembly to illustrate the geometry of a comb pair is presented in figure 3.25. The combs which are marked with a V (grey background) are on a high voltage potential, whereas the middle comb (white background) lays on ground potential. The distance between the comb is two time g and the dimensions of a comb teeth is $2c \times 2d$. The engagement distances of the comb pair is marked as $2x_0$.

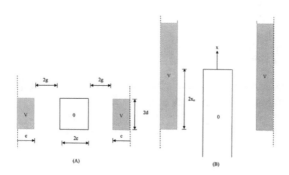

Figure 3.25: Schematic of a single comb pair of the ESC. A: End view and B: Side view. The voltage is applied to the shaded combs, the unshaded comb is electrically grounded [100]. The grounded comb is usually placed on the pendulum.

Johnsan and Warne [100] showed that for a comb configuration where the separation gap of the combs is equal to the comb dimensions (i.e. $g = c = d$) a simple analytical formula can be used to calculate the force that are produced by the combs, that is given by

$$F_{ESC} = 2 \cdot N \cdot \epsilon_0 \cdot V_{ESC}^2 \cdot \left(1.0245 - \frac{g}{\pi \cdot x_0}\right) \quad , \quad (3.7)$$

where N is the number of comb pairs, ϵ_0 is the permittivity of the gap medium (usually vacuum) and V_{ESC} is the potential difference between the combs. The equation is valid for semi infinite combs when the ratios of $x_0/g \geq 1.5$. For large ratios the force asymptotes near $2 \cdot N \cdot \epsilon_0 \cdot V_{ESC}^2$, i.e. the force is relatively independent over a reasonable range of pendulum deflection. This fact is the major advantage of an ESC compared to other electrostatic actuators. The comb assembly used has five comb pairs ($N = 5$), the gap between the combs is one millimetre wide ($g = c = d = 1\,\text{mm}$).

It is clearly evident, that a precise calibration is most important to enable highly precise thrust measurements. An experimental assessment of the real balance sensitivity allows a thrust measurement which is independent from manufacturing and integration tolerance. Moreover, it can be used to prove and extend the theoretical model that was used to design the balance. An error calculation is necessary to ensure the calibration is trustworthy. Assuming, the errors are normally distributed and the variables in equation 3.7 are defective and independent, the absolute error can be estimated as the sum of the first order Taylor series expansion for every single variable. Due to the propagation of uncertainty for a function ($f(x, y, z, ...)$), the absolute error can be written as

$$|\Delta X| = \sqrt{\left|\frac{\delta f}{\delta x}\right|^2 \cdot \Delta x^2 + \left|\frac{\delta f}{\delta y}\right|^2 \cdot \Delta y^2 + \left|\frac{\delta f}{\delta z}\right|^2 \cdot \Delta z^2 + ...} \quad , \qquad (3.8)$$

where Δx, Δy, Δz are the absolute errors of the single variables [102]. Hence, the theoretical equation for the absolute error of the ESC is given by

$$|\Delta X_{F_{ESC}}| = \sqrt{\left|\frac{\delta F_{ESC}(V, g_0, x_0)}{\delta V}\right|^2 \Delta V^2 \left|\frac{\delta F_{ESC}(V, g_0, x_0)}{\delta x_0}\right|^2 \Delta x_0^2}$$
$$+ \left|\frac{\delta F_{ESC}(V, g_0, x_0)}{\delta g_0}\right|^2 \Delta g_0^2 \quad . \tag{3.9}$$

The absolute error of the high voltage supply (ΔV) can be estimated with respect to the relative error of the power supply (Heinzinger LNC Series $\pm0.02\,\%$) and a typical calibration voltage (e.g. $1000\,\mathrm{V}$). Therefore, the absolute error of the voltage supply is assumed as $0.2\,\mathrm{V}$. The absolute error of the g-parameter can be estimated with respect to the manufacturing tolerances (ISO 2768, fine) as $\Delta g_0 = 0.05\,\mathrm{mm}$. The xyz-translation stage used has a specified sensitivity of $0.01\,\mathrm{mm}$. Under consideration of the presented values, the equations 3.9 and 3.7, the relative calibration ($f_{cali_{ana}}$) error becomes $1.2\,\%$. It must be underlined that this analytical error calculation considers only the specified error of power supply, the manufacturing tolerances and only one degree of freedom misalignment error, in the direction where the setup should be anyway insensitive and indeed in the present case this error is insignificant small ($\leq 0.1\,\%$).

It can be assumed that misalignment in other degrees of freedom must play a more important role in the error budget of the calibration. A test and detailed analysis with a slightly modified ESC assembly was performed by B. Seifert et. al. [45]. They have demonstrated that all degrees of freedom of the ESC play a role for the error calculation.

In order to avoid the limitations of the performed analytical error calculation, a numerical parameter study of the calibration device used has been performed to assess the influence of the misalignment in all degrees of freedom. As a result of the study a more precise calibration uncertainty can be estimated. The numerical simulation was performed in the COTS FEM software Maxwell 3D. In a first step, the numerical simulation was verified with the analytical equation given above (equation 3.7). For the whole simulation space ($0\,\mu N$ to $1000\,\mu N$), the numerical model and the analytical model were in agreement. It has been found that the highest calibration error is a result of an overlaid disorientation in all translation directions and rotation angles. The error is strongly dependent on the absolute magnitude of the values.

However, assuming that the maximum rotation uncertainty in all degrees of freedom is $\leq \pm 1.5°$ and the maximum translation error in all directions is $\leq \pm 0.125\,mm$ caused by pendulum motion and adjustment tolerance the relative error ($f_{cali_{num}}$) becomes $\leq \pm 3\,\%$, i.e. the relative error of the calibration in a range from $0\,\mu N$ to $1000\,\mu N$ is predicted as $\pm 3\,\%$. The whole dataset of the performed numerical simulation is available in appendix ??.

For a reliable, accurate and trustworthy thrust measurement an exclusive error calculation of the calibration is not sufficient because other random and systematic errors can falsify the measurement results. Therefore, an additional error analysis has been performed and possible balance parts that might play a role had been identified. To gain a complete overview which parameters are involved, the information which was presented in section 3.3.1 can be wrapped up. That means basically to unify the equations 3.2, 3.3, 3.5.

For the case of the open loop measurement, the measured force can be written as

$$F_{meas} = S_{calOL} \cdot \Delta x \quad ,$$

$$F_{meas} = \frac{l_{mirror}}{l_{thruster}} \cdot K_{pendulum} \cdot \Delta x \quad , \quad (3.10)$$

$$F_{meas} = \frac{l_{mirror}}{l_{thruster}} (\frac{E \cdot w_{total} \cdot t_{spring}^3}{6 \cdot l_{spring}^2 \cdot l_{thruster}} + \frac{m_p^2 \cdot g_0 \cdot l_{cog}}{I_{pendulum}}) \cdot \Delta x \quad .$$

The given equation is valid as long as the pendulums are almost physically equal. Considering the presented equations and the set of estimated input parameters the absolute error for a set of working points can be estimated. The estimated absolute error parameters are presented in table 3.7. The values are derived from table 3.5 based on conservative assumptions for the possible error sources. For example a maximum temperature variation of $10\,K$ was taken into account. The estimation of the variation of the CoG and the moment of inertia because of unsymmetrical temperature variations was performed via the Computer Aided Design (CAD) software used, including a margin of safety of 10. With respect to the presented values the maximum relative error (f_{ol}) is $\leq 0.76\,\%$.

Table 3.7: Summary of the estimated absolute uncertainties of the independent input parameters to determine the overall balance uncertainty.

Parameter	Specific Value
$\Delta l_{thruster}$ ($\Delta T = 10$ K)	0.0458 mm
Δl_{mirror} ($\Delta T = 10$ K)	0.0144375 mm
$\Delta l_{voicecoil}$ ($\Delta T = 10$ K)	0.03696 mm
Δw_{total} ($\Delta T = 10$ K)	$1.08 \cdot 10^{-3}$ mm
Δt_{spring} ($\Delta T = 10$ K)	$8.1 \cdot 10^{-6}$ mm
Δl_{spring} ($\Delta T = 10$ K)	$4.86 \cdot 10^{-4}$ mm
Δm_p (caused by propellant flow)	0.01 kg
Δl_{CoG} (caused by unsymmetrical temp variations)	1 mm
$\Delta I_{pendulum}$ (caused by unsymmetrical temp variations)	0.001 kg mm^2
$\delta \Delta x$(Interferometer resolution)	5 pm/$\sqrt{\text{Hz}}$
ΔI_{vc}(Keithley current source resolution)	10 pA
ΔC_{vc}(Voice coil specification)	$2 \cdot 10^{-9}$ N/A

In closed loop operation, it has to be considered that in addition to the uncertainty of the thrust measurement the uncertainty of the actuator has to be taken into account. The force that the actuator applies to the pendulum can be described as

$$F_{meas_{vc}} = C_{vc} \cdot I_{vc} \cdot \frac{l_{voicecoil}}{l_{thruster}} \quad , \tag{3.11}$$

where C_{vc} is the force sensitivity and I_{vc} is the input current of the voice coil. With reference of table 3.7 the equation can also be used to form the first order Taylor polynomial in order to calculate the uncertainty of the force actuator. For the presented value the maximum relative error is ≤ 0.02 %, i.e. the overall relative error in closed loop (f_{cl}) is ≤ 0.78 %.

Table 3.8 illustrates that the dominating term in the uncertainty budget is the relative error of the calibration, due to geometrical misalignment of the ESC in multiple degrees the freedom. In worst case it can be assumed that the maximum relative uncertainty is not larger than 3.78 %.

Table 3.8: Summary of the estimated relative uncertainties for the different measurement modi and the calibration.

Parameter	Specific Value
$f_{cali_{ana}}$	1.2 %
$f_{cali_{num}}$	3 %
f_{ol}	0.76 %
f_{cl}	0.78 %

3.3.5 Thrust Balance Characterisation, Testing and Analysis

Before real thrust measurements with an active thruster can be performed, it is necessary to characterise the thrust stand, in order to obtain the key parameters of the thrust balance, i.e. the determination of the pendulum sensitivity, noise floor, transfer function, eigenfrequency and other characteristics. If these steps succeed, the thrust measurement procedure (see figure 3.23) will follow. The whole characterisation must be performed with the full experimental setup, i.e. the thruster assembly, the calibration device, the gas connection, the harness and all other parts are mounted on the pendulum and all interconnections are attached. This is important since every modification of the pendulum assembly results in a change of the key parameters, due to the variation of the moment of inertia, the CoG and the pendulum mass. That means, a characterisation of the setup has to be performed after every major change of the setup, e.g. change of the thruster, adding, or removing devices which are mounted onto the pendulum, etc.

In the following section all major steps of the thrust stand characterisation will be presented and explained. As first step of the characterisation, the sensitivity of the thrust stand has to be determined. Thus, a calibration must be performed. Figure 3.26 presents a calibration run of the pendulum balance. The translation (in nm) of the balance (solid curve and scale on the left side) is plotted versus time. In comparison to this, the applied force (in µN, dashed curve and black scale on the right side) is plotted versus time, too. The calibration had been performed within 230 s. The force was applied via the ESC. Each step was 0.5 µN and 5 s long. A full calibration cycle is shown. The translation of the balance followed the applied force instantly. The difference between the rising shoulder of the curve and the falling shoulder of the curve is minimal and inside the approximated uncertainty of the calibration (≤ 3 % see section 3.3.4). Moreover, the figure illustrates the repeatability and the drift of the balance. At the displayed time scale, the drift is smaller than 0.1 µN.

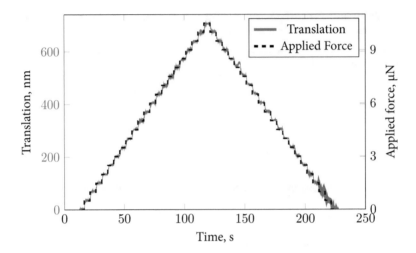

Figure 3.26: The figure presents a calibration of the balance. The balance translation (solid line) and the applied force (dashed line) are plotted versus time.

The raw data of the calibration that was given in figure 3.26 has to be processed in order to determine the balance sensitivity. Typically, the data points on each thrust level are used to calculate the average value for every step. The data points are plotted as force versus translation. An example is presented in figure 3.27. The force (in μN) is plotted versus the translation (in nm). The calibration run was also performed in 0.1 μN steps, from 0 μN to 80 μN. The upper (solid, red) scatter plot shows the result of the calibration for the measurement pendulum and the lower (solid) scatter plot presents the results for the reference pendulum. The slope of the scatter plots is equal to the spring rate. A linear interpolation was performed for each scatter plot to estimate the spring rates. For the interpolation a least square linear fit was used; the norm of residuals for the fits is smaller then 10^{-5} N/m. The measurement pendulum has a spring rate of 0.0157 μN/nm and the reference pendulum has a spring rate of 0.01812 μN/nm. In the presented case, the results and figure 3.27 indicate, that each pendulum has a different spring rate. This implies that the pendulums are not fully symmetric. A possible reason is that the fabrication and integration tolerances of the assembly causes differences in the weight and in the mass distribution of the pendulum. The example illustrates a typical behaviour of the first characterisation of a firstly tested balance setup, e.g. after changing the thruster under test. As result the pendulum has to be realigned with the counter weight. To verify the result, the calibration of both pendulums has to be re-performed.

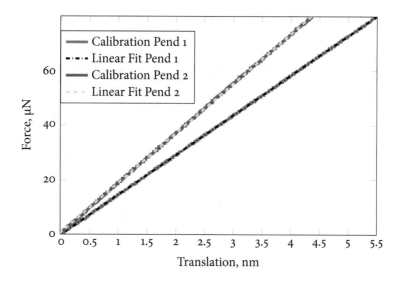

Figure 3.27: Example of a calibration of both pendulums. The force (in μN) is plotted versus the translation (in nm). The upper (red) scatter plot shows the result of the calibration for the measurement pendulum and the lower (blue) scatter plot presents the result for the reference pendulum. The presented case illustrates that the spring rates of the pendulums can be unequal and therefore a fine tuning via counter weight is required.

After the realignment of the setup additional calibration runs have been performed, the spring rate of both pendulums has been adapted to $0.0201 \pm 6 \cdot 10^{-4}$ μN /nm. The mass of each pendulum was 4.4 kg (incl. thruster).

In figure 3.28 the result of a long term calibration test is presented, where the calibration run number is plotted versus the measured sensitivity. The squares are the measurement points with attached error bars. For the measurement 155 calibrations runs had been performed. Every 20 minutes a new run was started so that the presented data covers the variation of the thrust stand sensitivity in a time frame of about 51.6 hours. The variation of the pendulum sensitivity is less than 1.5 % and the mean value (0.0201 μN/nm) is inside the estimated uncertainty budget. The measurement underlines the stability of the whole thrust balance assembly. Moreover, the repeatability of the defined calibration process has been demonstrated.

After the successful alignment of the pendulums and the determination of the pendulum sensitivity, long term noise measurements can be performed to gain information about the long term stability, eigenfrequencies of the setup and in

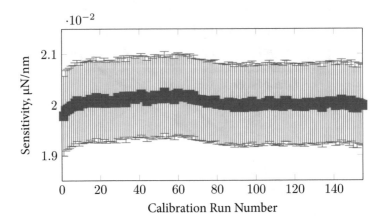

Figure 3.28: Result of a long term calibration test of about $51.6\,\mathrm{h}$ where the sensitivity is plotted against the calibration runs. The variation of the pendulum sensitivity is less than $1.5\,\%$.

general to specify the balance resolution in different frequency spectra. Moreover, the effects of different features of the thrust stand can be demonstrated.

Figure 3.29 presents the balance performance in different configurations. The shown PSD, where the thrust (in µN) is normalised to the frequency (in $\sqrt{\mathrm{Hz}}$) is plotted logarithmically, provides an overview of the noise level at specific frequencies. The dash-dot-dotted line plot presents the LISA requirement (see chapter 1), which is given by the required thrust level combined with the typical LISA allocation [7].

The dashed curve presents the noise measurement of the thrust balance, where a single undamped pendulum versus a fixed mirror was measured. Therefore, no common mode rejection occurs. The observed eigenfrequency of the pendulum is $0.77\,\mathrm{Hz}$, which is close to the theoretical estimated eigenfrequency of $0.81\,\mathrm{Hz}$. The amplitude of the swinging pendulum is the dominant noise term of the measurement resolution. At lower frequencies pink noise occurs which limits the resolution. Possible reasons are thermomechanical drifts and the movement of the whole setup. These movements are in the nanometre regime, which cause a permanent phase drift of the measurement signal.

The dotted curve presents the performance of one damped pendulum versus a fixed mirror (no common mode rejection). The damper suppresses the eigenfrequency. Therefore, at higher frequencies the noise level decreases 1.5 orders of magnitude. At lower frequencies pink noise dominates the PSD as shown in the dashed curve. The plot illustrates the function of the eddy current brake and why this sub-assembly is important for the overall thrust balance functionality. The

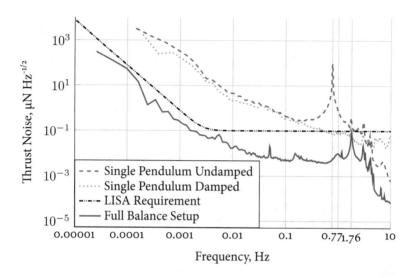

Figure 3.29: Summary of the balance performance. The dashed curve presents the performance of one undamped pendulum. The dotted curve shows the performance of
a single damped pendulum (no common mode rejection). The dash-dot-dotted
curve represents the LISA requirement. The solid curve illustrates the performance of the full balance setup with active common mode rejection. The thrust
balance fulfils the LISA thrust noise requirement.

eigenfrequency of the pendulum would directly be limiting the resolution inside
the targeted measurement bandwidth. Moreover and as observed in [48], the eigenfrequencies of the pendulums is continuously phase shifting, that means that a
common mode rejection would not be possible.

The noise measurement of the full setup, i.e. two damped pendulums, is shown
as the solid line plot. The noise level has been decreased by one order of magnitude.
The assembly is stable enough to fulfil the requirement. This also underlines the
result of the calibration measurement shown in figure 3.26. The curve demonstrates
that the balance can measure in sub-micro-Newton range and fulfils the required
stability to perform thrust noise characterisation of possible LISA AOCS thruster
candidates.

At higher frequencies the resolution is limited because of a peak at 1.76 Hz. Measurements with different sample frequencies make clear that the peak is not an aliasing
effect. However, it had been found that this peak is one of the eigenfrequencies of
the vacuum tank assembly which is floating on a nitrogen pillow. In front of the
peak, from 1.76 Hz to 10 Hz the curve presents a low-pass-filter behaviour. This

is reasonable, since the eigenfrequency is lower than this frequencies which also means that the amplitude of the signal is damped.

As last step of the thrust balance characterisation a measurement of the transfer function of the balance is performed. In parallel to the noise measurement, the transfer function provides information about the frequency dependences of the balance sensitivity. The response time can also be determined from the transfer function. To measure the transfer function, an active frequency sweep is used. The frequency sweep is applied either by the ESC, or by the voice coil. During the measurement the translation data of the balance is stored in parallel to the data of applied force.

Figure 3.30 presents one of the performed frequency sweep. In this case a voice coil was used to apply a force onto the balance. The input current (solid line, in nA) and the measured translation of the balance (dashed line, in nm) is plotted versus time (in s). For the data processing it was assumed that the input current is proportional to the applied force. The figure illustrates a full run. It started at a low frequency of 0.01 Hz. The frequency was logarithmically increased to 20 Hz, followed by a logarithmically decrease back to 0.01 Hz. After the measurement both data sets are used as input of a fast Fourier transformation to transpose the data into the frequency space. At the end of the process the Bode plot is generated.

The result of the performed measurement is presented in figure 3.31. The shown Bode plot presents the frequency (in Hz) versus the amplitude (in μN/nm) as the upper plot, called magnitude plot. The lower plot shows the phase plot where the frequency is plotted versus the phase of the signal. The curves illustrate a set of transfer function measurements. The dotted curves and the dash-dotted curves are the Bode plot of a thrust balance setup where the eddy current brake was set to a lower influence by extending the distance of the permanent magnets to the conducing plates. Therefore, the eigenfrequency is not fully damped. It is visible at 0.7 Hz.

The black solid curves present a measurement with fully activated eddy current brake. It can be observed that the natural frequency of the thrust stand is fully damped, the system is overdamped. Therefore, the response time of the pendulum is much lower, and the sensitivity at higher frequencies is reduced. The effect is also evident in the phase plot. The higher damping leads to a high phase shift which also would influence the PID control loop of the voice coil in closed loop operation.

To demonstrate that the assessment of the Bode plot is independent from the actuator that is used to apply the frequency sweep on the thrust balance, the first measurement was performed with two different actuators. The frequency sweep to generate the dotted curve was performed with the ESC, whereas the frequency sweep of the dash-dotted curve was performed with the voice coil. It is clearly evident that

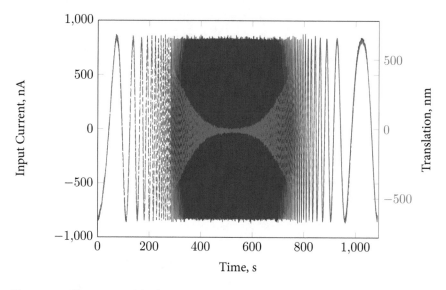

Figure 3.30: Illustration of the frequency sweep used. The input current of the voice coil
(solid line, in nA) and the measured translation of the balance (dashed line, in
nm) is plotted versus time (in s).

no difference between the curves can be observed, i.e. the actuator has no influence
on the measurement result.

By connecting the result of the noise measurement which was presented in figure 3.29 and the transfer function measurement, it becomes clear that for a thrust
noise measurement that should provide information above the eigenfrequency, a
correction of the thrust noise measurement has to be performed. The Bode plot can
directly be used for the correction of the PSD, since the Bode plot data is the same
convolution as the PSD.

Figure 3.32 provides an example of the effects of the PSD correction. The plot
presents two PSDs. The frequency is plotted versus µN normalised to the frequency
(by $\sqrt{\text{Hz}}$. The dash-dot-dotted curve illustrates the LISA requirements. The solid
curve is the uncorrected PSD which was directly processed from the measurement
data with a Blackman Harris 92 window [10]. The uncorrected red curve is below
the requirement inside the whole bandwidth. This would mean that the result is
fully compliant. The dotted curve presents the correction of the solid curve via a
multiplication with the Bode magnitude plot. As expected, at low frequencies the
corrected and the uncorrected curves are equal, but at higher frequencies the noise
floor increases due to the correction. Especially at 8 Hz, the corrected curve is not

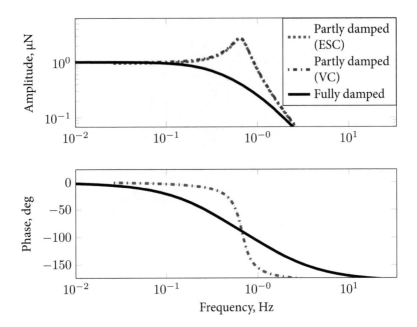

Figure 3.31: Example of the measured Bode plot of the thrust balance for different eddy
current brake configuration. The measured response were applied with different actuators to demonstrate that the measurement is not dependent on the
actuator used.

compliant with the requirement. The example underlines that for a trustworthy
thrust noise measurements, the transfer function of the balance has to be known
and in case the measurement bandwidth is above the eigenfrequency of the setup, a
correction of the thrust noise measurement data has to be performed in order to
generate reliable measurement data.

3.4 Test Facility Real Thruster Measurement Results and Analysis

Up to now, the test facility was used to perform various tests with different kinds
of thrusters. Mainly test campaigns with the in-house developed μHEMPTs have
been performed, but also a test campaign with the RITμX in cooperation with
Ariane Group in Lampoldshausen. The characterisation of the different μHEMPT

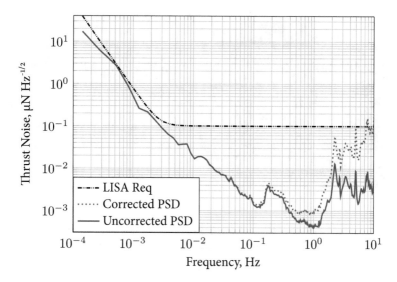

Figure 3.32: Example of the post processing of the measured balance noise floor to take the transfer function into account. The red curve illustrates the measured unprocessed PSD, whereas the blue curve shows the processed PSD. Due to the transfer function of the thrust balance the true PSD has a slightly higher noise floor.

laboratory models will be presented and analysed in chapter 4, while the performed tests with the RITµX will be presented in this section.

The goals of the RITµX testing in the Airbus facility was to demonstrate the unique capabilities of the thruster, especially the low thrust noise of the RIT technology. Simultaneously, the test campaign was used to prove and to practically demonstrate the ability of the test facility to perform a precise and reliable thruster characterisation with an already well characterised and understood thruster. For example, a similar thruster was tested at Estec [103] and Gießen [34]. It was also tried before to perform direct thrust noise measurements with a similar thruster [104]. Moreover, the thruster had performed an endurance test at ESTEC in 2015 [105].

The engineering model of the RIT µX is presented in figure 3.33. On the right side a picture of the thruster is shown and on the left side a picture of the same thruster is presented during firing in Airbus Friedrichshafen vacuum facility. In both pictures the acceleration grid can be recognised. The thruster has an operation range between 50 µN to 2500 µN, while is applicable to the requirements of Euclid and NGGM.

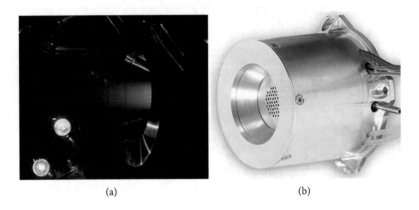

(a)   (b)

Figure 3.33: Photography of the characterised RITμX . In figure (a) the thruster firing is
presented whereas figure (b) shows the thruster outside of the vacuum chamber.

All required components (RFG, power supplies, controller) that are required to
operate the thruster are designed to fulfil the requirements of the missions mentioned.
As presented in figure 3.1,3.14 and 3.16, the RFG was also placed on the pendulum.
To cool the RFG a radiator was mounted on top of it. The thruster system was
consisting of the thruster, the RFG, a high voltage power supply for the screen grid, a
high voltage power supply for the acceleration grid, a power supply for the RFG and
a mass flow controller. All parts of the thruster system except the flow controllers
were provided by Ariane Group Lampoldshausen.

After the successful integration of the thruster, firstly a commissioning of the
thruster system was performed confirming the following: The communication
with the laboratory flow controller used was working properly, the communication
between the thruster control unit and the facility was working and the background
pressure inside the vacuum tank was below a sufficient level and in agreement with
the previously performed estimation during thruster firing. The pressure was in the
range of $1 \cdot 10^{-6}$ mbar and $3 \cdot 10^{-6}$ mbar when the thruster was operating. Also
temperature measurements were initially performed to ensure that the balance and
the thruster assembly were not exceeding their thermal specification. Unfortunately,
the radiator system was not able to dissipate the power loss over the whole operation
range of the thruster. Hence, the thruster was only operated at its lower power and
thrust levels during the test campaign.

Firstly, measurements with the plasma diagnostic were performed as part of
the commissioning. A representative result is presented in figure 3.34. The curve
illustrates a full Faraday cup plume characterisation. The angle (in deg) is plotted

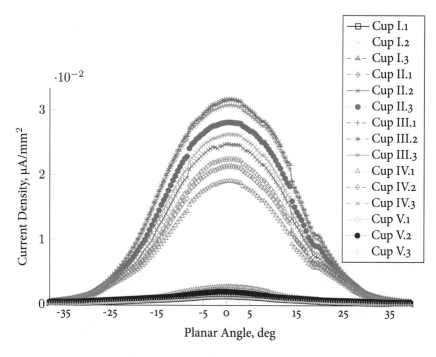

Figure 3.34: Result of the RITµX plume characterisation. The measured current density is plotted versus the planar angle. Each curve represents a single Faraday cup.

against the current density (in $\mu A/mm^2$). The measurement result of each cup is presented as a separate curve. The jib arm performed 0.5° steps. At every step 100 sampling points were taken. The measurement points shown are the mean value of these samples. With respect to the expected high divergence efficiency of the thruster the Faraday cups were arranged as close as possible to increase the lateral resolution of the measurement. Qualitatively, the plotted curves appear as expected. The plume has a Gaussian shape. Also the low current densities of cup array I and cup array V are in good agreement with the other Faraday cup measurements because the cups of cup array I and V are placed at 30°, i.e. β_i for cup I is between $-28.5° - -31.5°$ and for cup V 28.5° - 31.5°. The Gaussian shape of the plume is only disturbed by a small fall off around 15°, this is the region where the neutraliser is placed. For the neutralisation of the ion beam simple filament lamps as hot cathode were used. This effect was previously observed with the thruster and a comparable neutraliser at tests at the University of Gießen. However, the divergence efficiency of the thruster could be determined as $\geq 95\%$.

The result of the integration of the current density curves presented was 1.808 mA for the whole ion beam. The measured beam current was 1.8 mA, i.e. the result of the Faraday cup measurement is almost equal to the set value of the beam current controller.

This result is in agreement with the measurements mentioned that were previously performed with the thruster. Therefore, this measurement can be seen as the proof that the developed plasma diagnostic system works as expected.

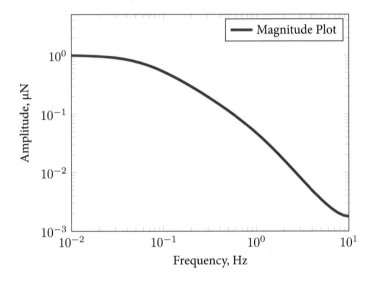

Figure 3.35: Transfer function of the balance configuration used. During the measurement campaign the balance was highly overdamped but due to time constraints it was not possible to adapt the setup to increase the sensitivity to higher frequencies.

Before the first thrust measurements were performed a full characterisation of the pendulum setup was undertaken. The characterisation followed the standard sequence presented (see section 3.3.5). Firstly, the sensitivity of the thrust stand was measured via a calibration and determined as to be 0.025 μN/nm. Secondly, a noise measurement was performed and the transfer function of the thrust balance was identified. The result is presented as magnitude plot in figure 3.35. The curve shows that the assembly was highly overdamped. Normally, a re-adjustment of the eddy current brake would be performed, but due to a challenging measurement schedule this was not possible. However, for the scheduled measurements the slow response time was not a limiting factor.

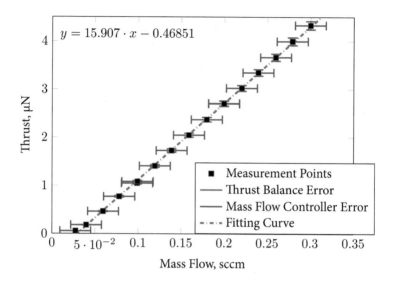

Figure 3.36: Result of the characterisation of the neutral gas thrust. The thrust measured is presented against the mass flow measured.

For highly precise electric propulsion testing, an assessment of the thrust generated from neutral particles is required, since the typical thruster control mechanism does not consider the so-called neutral gas thrust. To gain information about the amount of neutral gas thrust that is generated by the thruster, a characterisation of the influence of the neutral gas flow on the commanded thrust was made. The tests are summarised in figure 3.36. The plot presents the mass flow (in sccm) versus the measured thrust (in µN). As mass flow controller a Bronkhorst El-flow F-200CV was used. The squares are the measurement points. Every point is the mean value of a 90 s long measurement step, or in other words the mean value of 1800 data points. The horizontal bar is the absolute error of the mass flow meter used. The vertical bar represents the error of the thrust balance. At 0.1 sccm and 0.2 sccm two measurement points were taken. The first point at the beginning of the measurement run the second point at the end of the overall measurement run to demonstrate that the measurement generates reproducible results. As presented in figure 3.35 the points are almost totally overlapping and in agreement with the illustrated uncertainty bars.

To estimate the thrust produced by the mass flow, a linear fitting was performed. The dash-dotted line represents the estimated fit. For all data points, the fit is inside the estimated errors. Therefore, a linear correlation between the mass flow and the neutral gas thrust is a valid assumption. But a notable offset of the presented fit can be observed. The most probable reason for the offset is that the valve of the flow controller, or the integrated flow meter, generates this offset. Another reason is that the free molecular flow at low pressure or mass flow levels causes a non-directional movement of the particles. Therefore, no direct force is produced. A leakage of the tubing used can be excluded because of helium leakage tests which have been performed. During the measurement presented the neutraliser and the thruster power suppliers were turned off.

The result of the linear fit performed is presented at the top left side of the chart. The formula was used to estimate the impact of the neutral gas thrust in the following measurements.

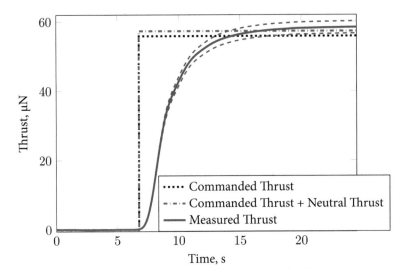

Figure 3.37: Example of a thrust measurement performed which illustrates that the neutral gas thrust can influence the total thrust of the thruster. The measured thrust is plotted versus time. The dotted curve presents the commanded thrust where the neutral gas thrust was not considered. The dash-dotted curve illustrates the commanded thrust plus the estimated thrust caused by neutrals.

An example of the uncertainty that is generated due to neutral gas thrust in micro-Newton electric propulsion is given in figure 3.37. The plot presents a thrust measurement from the deactivated thruster to 58 μN, where the time in seconds is plotted versus the thrust in micro-Newtons. The commanded thrust is illustrated as dotted line, the measured thrust is shown as solid line and the error of the thrust measurement is given as dashed line. If only the commanded thrust would be taken into account the thrust measurement would have a difference of more than 4.5 % to the commanded value. But, if we take into account that the thruster has a low mass utilisation efficiency of about 20 %, at the operation point used, it becomes obvious that the thrust of neutral gas flow has to be considered.

Hence, the predicated neutral gas thrust can be added to the commanded thrust which was performed in the dash-dotted curve. The neutral gas thrust was estimated with the overall propellant mass flow and the equation from figure 3.36 to be 0.93 µN. The expected thrust value is now inside the uncertainty of the thrust measurement and the difference between the measured and the predicted value is ≤ 1.5 %.

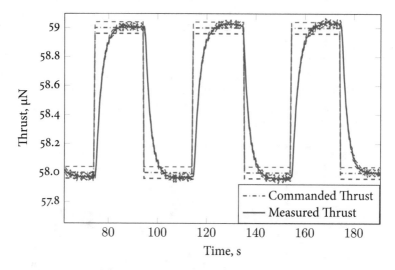

Figure 3.38: A 1 µN stepping performed by the RITµX on an absolute thrust level of 58 µN. The thrust commanded is shown as dash-dotted line, the measured thrust is presented as solid curve. The error bars of both data sets is given as dashed curves.

As previously explained, the goal of the test campaign was to demonstrate the ability of the thruster to perform highly precise pointing, as well as the thrust balance is able to measure small thrust steps which are generated by a real thruster. Therefore, different tests were performed to validate these capability. Figure 3.38 presents a measured 1 µN stepping and figure 3.39 shows a 0.1 µN.

In figure 3.39 a 1 µN stepping of the thruster is presented, where the thrust (in µN) is plotted versus the time (in s). The stepping started at 58 µN. During the stepping the mass flow and the grid voltages were kept constant. The fine controlling of the thrust was exclusively performed by alternating the RFG power; this control loop is called beam current controller. As actual value the beam current was used. The ampere-meters which were part of the power supplies used were the limiting factors inside of the control loop. The resolution of the ampere-meters were 0.001 mA that corresponds to 0.0404 μN at the specific operation point. The resolution of the ampere-meter, or the range of the power supplies, is a consequence of the wide operation range of the thruster. The blue curve illustrates the commanded thrust and the dashed blue lines present the uncertainty of the beam current control loop.

The solid curve represents the measured thrust. The dashed-dotted line presents the commanded thrust. The dashed line represents the uncertainties of the measurement. It can be observed that the measured thrust follows the commanded thrust directly, with respect to the transfer function of the thrust stand. The measured thrust is inside the uncertainties.

The plot demonstrates the ability of the thruster to perform precisely micro-Newton stepping which is required for the targeted future scientific space missions such as NGGM and Euclid. Moreover, it underlines the capability of the thrust stand to measure these steps. The curve also illustrates that the uncertainty analysis performed is in agreement with the measurement results.

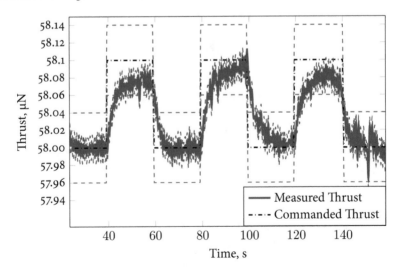

Figure 3.39: The Figure presents a 0.1 μN stepping performed by the RITμX on an absolute thrust level of 58 μN. The commanded thrust is shown as dash-dotted black line, the measured thrust is presented as solid red line. The error bars are plotted as dashed lines. All curves are plotted versus time.

The thrust stepping was performed at various thrust levels and with different step sizes. The smallest step size that was tested was $0.1\,\mu N$ which is presented in figure 3.39. Test stepping were performed on an absolute thrust level of $58\,\mu N$. The thrust (in μN) is again plotted versus the time (in s). The dash-dotted curve presents the commanded thrust and the dashed curve illustrates the absolute error of the commanded thrust. The solid curve represents the thrust which was measured by the micro-Newton thrust balance. The dotted red line represents the absolute error of the direct thrust measurement. The measured thrust is inside the estimated error interval.

Figure 3.39 underlines the ability of the thrust balance to measure in a sub micro-Newton regime. It verifies the previously presented measurement results and confirms the ability of the thrust balance to characterise the thrusters in the targeted measurement range that is required for LISA and other future scientific space missions.

Furthermore, the figures demonstrate the ability of the RITµX tested to generate sub micro-Newton steps. This underlines the capability of the thruster for highly precise AOCS with a resolution smaller than $1\,\mu N$ or even smaller forces, although the thruster system was not specifically designed to perform stepping in the sub micro-Newton regime.

As mentioned, the thrust balance noise characteristic was tested before the first thrust measurements were carried out. Thrust noise measurements of the RITµX were also performed. The PSD shown in figure 3.40 presents the results of the thrust noise measurements performed. All PSDs presented have been corrected with the measured and previously shown transfer function of the setup in order to present the real thrust noise performance of the characterised devices. The thrust, which is normalised to the frequency in $\mu N/\sqrt{Hz}$, is logarithmically plotted against the frequency in Hz. The PSD provides an overview of the noise level at specific frequencies.

The Euclid AOCS and the NGGM thrust noise requirements are plotted as a single dashed curve. As mentioned, this is of special importance because the RITµX is considered as a suitable AOCS thruster design for the NGGM spacecraft.

The dash-dot-dotted curve presents the LISA requirement (see chapter 1), which is given by the required thrust level combined with the typical LISA allocation. The performance of the thrust balance is shown as solid curve. The solid PSD was generated from a 20 h measurement, with a Blackman-Harris 92 window [10]. A linear de-trend was also performed.

The dotted curve presents the performance of the RITµX. To generate the PSD, an 11 h long measurement and a Blackman-Harris 92 window [10] with linear de-trend was used. At frequencies between 1 Hz and $8 \cdot 10^{-3}$ Hz the thrust noise of the thruster is below $0.1\,\mu N/\sqrt{Hz}$. This can also be observed in the time domain data

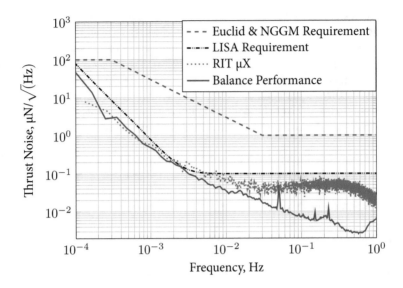

Figure 3.40: The PSDs provide an overview of the noise levels at specific frequencies.
The logarithmically plotted thrust, which is normalised to the frequency in
$\mu N/\sqrt{Hz}$, is plotted against the frequency in Hz. The dashed and the dash-
dot-dotted curves represent the considered requirements, the dotted curve
illustrates the thruster performance. The solid curve presents the balance
performance. The dotted curve illustrates that the RITµX almost fulfils all
requirements apart from a small frequency band, although the thruster was not
optimised for LISA.

which are presented in figures 3.38 and 3.39. With respect to the error budget of the
thruster assembly, the noise level of the thruster is defined by the power supplies
used and in particular by the resolution mentioned (0.001 mA, or 0.0404 µN at the
specific operation point) of the ampere-meters used for the beam current controller.

The thrust noise level increases at frequencies below $8 \cdot 10^{-3}$ Hz typically with
a $1/f$ shape caused by thermal drifting of the electronics used, as well as thermomechanical drifting. Below $1 \cdot 10^{-3}$ Hz the curve of the thruster noise level is almost
equal to the thrust balance noise level curve, i.e. at these measurement points no
statements of the thruster performance can be given.

The measured thruster demonstrated over the whole measurement bandwidth a
noise level which is more than one order of magnitude below the NGGM & Euclid
requirement, for which it is designed. Moreover, the thruster almost fulfilled the
LISA requirement. Only between $2 \cdot 10^{-3}$ Hz and $8 \cdot 10^{-3}$ Hz the noise level was

above the requirement. Considering, the limitations of the power supplies used it seems to be feasible to fulfil the requirement with the use of electronics that are specially designed for the LISA mission application.

Table 3.9: Summary of the measurement key parameters of the test campaign.

Parameter	Value
Tested thruster weight	$\geq 1\,\text{kg}$
Max power on balance	$33\,\text{W}$
Thrust Noise	$\leq 0.1\,\mu\text{N}$ (1 Hz to $8 \cdot 10^{-3}$ Hz)
Demonstrated Thrust Steps	$0.1\,\mu\text{N}$
Divergence Efficiency	$\geq 90\,\%$

In summary, we can say that the measurements performed confirmed the targeted capabilities of the micro-Newton thruster test facility to characterise electric micro-Newton thrusters in the whole LISA bandwidth. Furthermore, the facility is able to determine the key parameters of an electric thruster. The result of the test campaign is summarised in table 3.9. Moreover, the experimental characterisation validated the uncertainty estimation performed due to the fact that all measurements were inside the predicted error budget.

4 Micro High Efficiency Multistage Plasma Thruster Development

As mentioned in the introduction, Airbus started the development of the micro-Newton HEMPT in 2009 with the goal to demonstrate that the HEMPT technology could be used for highly precise AOCS. Until that time various laboratory μHEMPT had been built and tested.

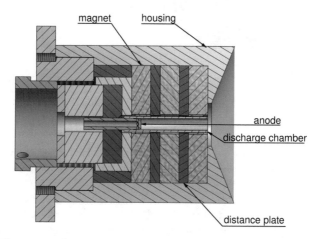

Figure 4.1: Illustration of a cross sectional snap shot of the EBB20 that was the starting point of the thruster development performed [32].

The geometrical variation survey of A. Keller, where more than 25 different thruster variations had been tested between 2008 and 2013 [32], representing the precursor of the HEMPT development presented in this thesis. Hence, a summary of the design and the thruster performance achieved until 2013 is presented in this chapter, followed by the presentation of the current thruster development performed.

A representative thruster design which has been used during the study is illustrated in figure 4.1. The cross section of the thruster shown (designated as EBB20) presents the magnet stack that consists of three magnets, the discharge chamber, the anode and the other structure. Between the single magnets copper spacer are placed.

© Springer Fachmedien Wiesbaden GmbH 2018
F. G. Hey, *Micro Newton Thruster Development*,
https://doi.org/10.1007/978-3-658-21209-4_4

During the study, the distance of the magnets was varied between 1 mm and 10 mm. In the centre of the magnets, the discharge chamber is placed, it is made of Alumina. The inner diameter of the discharge chamber was varied as well in the range between 2 mm and 8 mm. The thruster housing surrounds and supports the magnet assembly.

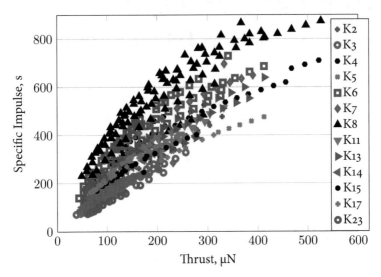

Figure 4.2: Summary of the thruster performances achieved at the end of the performed parameter study of A. Keller [32]. The plot presents the specific impulse versus the indirect measured thrust of different thruster variations.

Figure 4.2 summarises the thruster performances obtained between 2008 and 2013. The chart presents the thrust (in μN) versus the specific impulse (in s) for different μHEMPT versions. Every data point represents an operation point tested of the specific thruster. In general the plot demonstrates that some of the thruster variations were able to produce thrust in the micro-Newton regime. However, the thruster operation envelope presented illustrates that the maximum specific impulse was not higher than 850 s and at the low thrust operation point the specific impulse rapidly decreased.

There were several reasons for the low specific impulses that were achieved. For example during the A. Kellers study it has been demonstrated that the divergence efficiency of the thrusters tested were below 55 %. Figure 4.3 presents a representative ion current density measurement which has been performed during the survey to illustrate the distribution of ion current inside the thruster plume. The chart presents the measured ion current density (in μA/mm^2) versus the planar angle (in

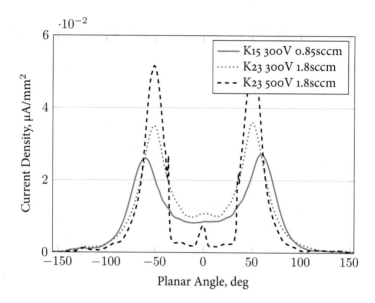

Figure 4.3: The plot illustrates the shapes of the plume of different EBB20 thrusters. For each configuration the ion current maximum is at angles wider than 50° which leads to high divergence loss.

deg). The curves enable a comparison of two different thruster variations (solid line versus dashed and dotted curves) and a comparison of the impact of the anode voltage at a specific operation point. In all curves the divergence of the beam is larger than 50 %. The difference between the thruster types presented is around 50 %. The curves illustrate that beside the high divergence of the whole ion beam also the distribution of the ions inside the plume are focused to the outer angles which additionally reduces the divergence efficiency (see equation 2.26).

The determination of the acceleration efficiency was also part of the geometrical variation survey. For the thruster tested an acceleration efficiency of typically 90 % could be demonstrated.

The thrusters were usually operated with mass flow levels above 0.5 sccm which results in the a reduced specific impulse in low thrust operation.

The result of the parameter study in comparison with the goals of the hereby presented thruster development which were derived from the LISA requirements are given in table 4.1.

The table illustrates that it is generally possible to operate a HEMPT in the μ-Newton regime and that also the acceleration efficiencies of the tested thrusters were promising. However, the specific impulse at low thrust levels and the low divergence efficiency were not sufficient for an EP thruster. Especially, the specific impulse should be greater than 1000 s to justify the use of such a thruster system, due to the higher system complexity compared with cold gas micro-thruster systems.

To overcome these limitations and to demonstrate that the HEMPT principle can be used for highly precise AOCS, a new development approach has been chosen that do not exclusively vary the geometrical thruster parameters in order to identify thruster configurations which have an improved performance.

Table 4.1: Comparison of the µHEMPT performance achieved until 2013 and the goals of the recent thruster development.

Parameter	Best Value until 2013	Targeted Value
Minimum thrust	$\leq 50\,\mu N$	$\leq 50\,\mu N$
I_{sp} in low thrust	300 s	$\geq 1000\,s$
PTTR	30 W/mN	30 W/mN
Divergence efficiency	$\leq 55\,\%$	$\geq 85\,\%$
Total efficiency	n/a	$\geq 20\,\%$
Thrust noise	n/a	$\leq 0.1\,\mu N\,/\sqrt{Hz}$

4.1 Next Generation µHEMPT Development Approach

As presented, until 2013 a geometrical survey of different µHEMPTs has been performed. Due to the high amount of measured data, the small difference and the coupling of the key parameters of the geometrical variations, it was challenging to extrapolate the optimal values of each geometrical parameter. Based on an analysis of the obtained data set the development strategy has been changed to a semi-empirical approach.

The semi-empirical approach includes a qualitative analysis of the magnetic field geometry of the thruster in comparison with the published basics of the HEMPT principle such as electron confinement in the cusp regions and a qualitative estimation of the key parameters of the electron motion according to the equations 2.14, 2.15 and 5. The estimations were supported by 2D and 3D numerical magnetic field simulations performed with the COTS tools FEMM and Ansys Maxwell 3D. Simple particle tracing was also performed with the COTS tool SIMION.

The principle was chosen to reduce the amount of data which has to be analysed to reduce analysis time and the required hot thruster tests which leads to an increased speed of the thruster development. This strategy includes also that only major thruster variations have been deeply analysed. Minor changes were only been characterised by the thruster operation parameters and of course cross checked if the specific minor change had an impact on the thruster performance. In this special case, minor changes were classified as variations of the thrusters that should have no impact on the thruster performance, such as mechanical modifications, or modification on the propellant feeding line etc.

The approach was also supported by numerical particle in cell simulations performed and published by Günter Kornfeld [71] and Tim Brandt [106, 107, 72].

As a first step the magnetic field topology of the previously tested thruster was analysed and compared with the general ideas of the HEMPT concept. To illustrate the general analysis procedure, a representative 2D magnetic field topology is shown in figure 4.4. The figure presents the result of rotation symmetry Finite Element Method (FEM) magnetic field simulation; the rotation axis is at the top side of the figure and the thruster exit is at the right side. In this case the magnet assembly is similar to the setup that is presented in figure 4.1. The three magnets are highlighted with a blue background. The ceramic discharge chamber is shown in white. The four cusp regions are marked with dashed circles. The separatrix of the exit cusp is marked by an arrow. The distribution of the magnetic field strength is underlaid as colour bar. Bright blue represents a small magnetic field strength whereas purple represents higher field strengths.

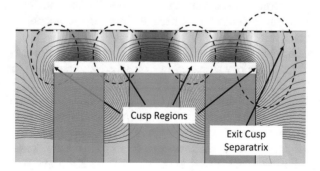

Figure 4.4: Illustration of the B-field configuration of the µHEMPTs tested until 2013. The black lines are the magnetic field lines, the coloured background illustrates the flux density of the B-field, bright blue is equal to very low flux density whereas purple presents high flux densities.

As presented in section 2.3 the cusp should confine the electrons inside the discharge chamber to increase the probability that an electron can ionise a neutral particle, which requires a magnetic mirror between the cusp. As presented in section 2.1, the magnetic mirroring requires a field gradient from a low magnetic field strength to a high magnetic field strength.

In contrast it can be observed that the field strengths gradient inside the cusps is low whereas in between the cusps the field strength increases. The configuration is not the same as the usual HEMPT magnetic field topology that has been published. This configuration does not reflect the electrons in front of the ceramic discharge chamber. Hence, the electron confinement is not optimal and this leads to additional energy loses and therefore to a reduced overall thruster performance.

In addition, to the unusual magnetic field topology the dimensions of the discharge chamber also play an important role during the thruster downscaling.

The maximum magnetic field strength is limited by the physical properties of the magnet material and therefore the maximum field strength of the magnets does not change during the downscaling. The Larmor-radii are also kept constant, which decreases the quality of the electron confinement when scaling the thruster system to smaller dimensions. The specific diameter of the Larmor-radii is also the physical limit of the HEMPT downscaling. According to equation 2.11 the Larmor-radii in the inside of the thruster is typically in the range of some tens of µm.

In contrast to, the linear downscaling of the magnetic field topology and strength, the inner diameter of the discharge chamber cannot be similarly down scaled because of the physical properties of the ceramic wall material. The wall thickness can only be down scaled to a certain extent. Thus, the maximum field strength and field gradient is significantly reduced due to the higher ratio between the chamber diameter and the chamber wall thickness.

The magnetic field topology has also an influence on the divergence efficiency. In particular, the field topology at the exit cusp has a critical impact on the divergence efficiency of the thruster plume [66]. The shape of the separatrix of the exit has to be optimised to achieve high divergence efficiencies. It is assumed that the electrons which are confined at the thruster exit are travelling outside along the separatrix surface, forming there a coaxial ring of electrons. Thus, a curved exit separatrix leads to a curved electron cloud at the thruster exit. The ions which leave the thruster are deflected due to the negative space charge of these electrons. Hence, a curved separatrix leads to high angular distribution of the thruster plume and therefore to low divergence efficiencies.

In 2006, TED published different concepts to create a flat separatrix which would lead to a better divergence efficiency [66].

Before the design of a completely new thruster generation was started, the results of the qualitative analysis of the magnetic field topology were validated with a modified thruster design from the previously performed parameter study, called EBB20 MkII. A cross section of the modified thruster is shown in figure 4.5. The thruster consists of the parts that were still available from the parameter study. A

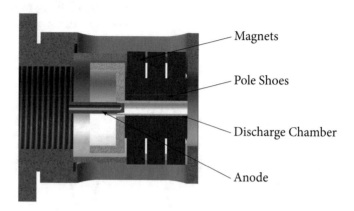

Figure 4.5: Cross section of the modified EBB20 (called EBB20 MkII) which has been
used to verify the empirically gained information about an improved electron
confinement. The field strength and the field gradient inside the cusp were
improved via a set of pole shoes between the permanent magnets.

stack of SmCo magnets surrounds an Alumina discharge chamber and a brass anode
is used. The inner diameter of the discharge chamber is 3 mm.

As mentioned, the small magnetic field strength gradient inside the cusp has been
identified as a major performance limiting factor of the old thruster design. To
optimise the magnetic field topology inside the cusp regions, pole shoes were used
instead of the copper spacer.

The effect of the modified magnetic field topology is illustrated in figure 4.6, where
the normalised thruster length is plotted versus the flux density. The curve presents
the variation of the field strength (in T) at the surface plane of the inner discharge
chamber wall. The cusps are located at position $0, 0.32, 0.66$ and 1 of the normalised
thruster length.

The dash-dotted plot (bottom) illustrates the flux density distribution in an unmodified EBB20, similar to the figure 4.4 the field strength at the cusps is minimal.
The dashed curve shows the field strength distribution in the case that no spacer
is placed between the magnets. In this case the minimum field strength inside the
cusp would rise up to 0.42 T, instead of 0.19 T.

The dotted curve presents the flux density distribution for the case if pole shoes
are used instead of copper spacers, similar to the EBB20 MkII configuration that
is shown in 4.5. Due to the pole shoes the field strength is concentrated inside the
cusp and the flux density rises up to 0.61 T.

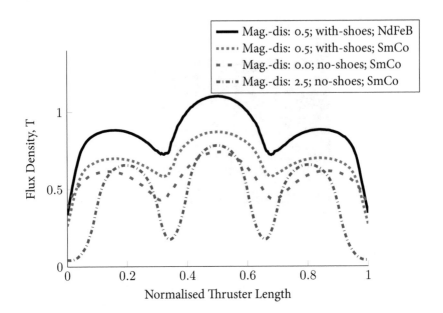

Figure 4.6: Comparison of the magnetic flux density in the plane of the discharge chamber. The flux density is plotted versus the normalised thruster length. The cusps are located at position 0; 0.33; 0.66; 1. The standard EBB20 configuration is shown as dash-dotted line (bottom), whereas the solid curve presents the B-field geometry of the EBB20 MkII with NdFeB magnets instead of SmCo.

A further enhancement of the overall magnetic flux density can be achieved by using NdFeB magnets instead of SmCo magnets because this type of rare-earth magnets is able to provide a significantly higher total flux density. The effect is illustrated as the solid curve; the flux density can be raised to 0.78 T, with the same magnet and pole shoe configuration as in the green curve and in the EBB20 MkII.

The NdFeB-magnets inserted offer a 12 % higher flux density. However, the Curie temperature of the NdFeB-magnets is specified as 180 °C instead of 350 °C for the previously used SmCo-magnets. In the case that the magnet stack reaches the Curie temperature an irreversible reduction of flux density occurs [110]. Additionally, the maximal theoretical flux density is dependent on the magnet temperature. Typically, the flux density decreases linearly if the environmental temperature rises. This decrease is usually reversible. Close to the Curie temperature the flux density starts to drop rapidly until it irreversibly reaches zero due to normalisation processes of the lattice. Figure 4.7 presents the general behaviour of the magnet alloys used at different temperatures [108, 109, 110]. The chart presents the theoretical maximum

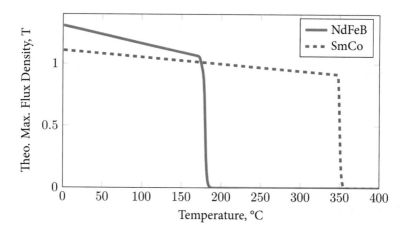

Figure 4.7: Illustration of the reversible and irreversible variation of the maximal magnet flux density. The solid curve represents the NdFeB alloy used, the dashed curve shows the SmCo alloy used. The theoretical maximal flux density (in T) is plotted versus the temperature [108, 109].

flux density (in T) versus the temperature (in °C). The solid curve illustrates the characteristic of the NdFeB-magnet alloy used whereas the dashed curve represents the characteristics of the SmCo-magnet alloy used.

In the inside of the operation space of the NdFeB-magnet, the flux density is higher than the flux density of the SmCo-magnet. However, due to the higher Curie temperature the SmCo-magnet offers a wider operation range than the NdFeB-magnet. It becomes clear that the advantages of the NdFeB-magnet can only be used if the thruster design allows a low temperature and thus a low power operation.

It must be mentioned, that for space applications also other material properties should be considered such as radiation hardness or mechanical characteristics. But due to the early development stages of the thrusters designed, these considerations were not made in the scope of this thesis.

The modified thruster was tested inside the Airbus micro-Newton thruster test facility. The result of the tests in comparison with the unmodified EBB20 is presented in table 4.2. Both thrusters had been operated at the last stable operation with a minimum propellant mass flow.

The EBB20 MkII could be operated until a minimum propellant flow of 0.26 sccm which is a propellant mass reduction of 52 % in comparison with the standard EBB20. The other operation parameters could be varied in the same manner as before. The reduced propellant mass flow leads to an increased specific impulse.

Table 4.2: Comparison between the two EBB20 thruster configurations at the operation point with minimal propellant mass flow.

Parameter	EBB 20	EBB 20 MkII
Anode Voltage	400 V	530 V
Anode Current	8 mA	6 mA
Mass flow	0.55 sccm	0.26 sccm
Calculated I_{sp}	360 s	710 s

As expected, the divergence efficiency was not improved due to the modified magnet field topology, because only the electron confinement inside the thruster had been optimised.

The measured data has shown that with the chosen semi-empirical approach serious thruster improvements can be made within a relatively short amount of time. Thus, the preliminary assessment validated the chosen approach and allowed a continuation of the activities with the target to develop a next generation µHEMPT which should overcome the performance limitations of the previously developed thrusters.

4.2 The Next Generation µHEMPT

The result of the preliminary test performed underlines that it is possible to generate valid assumption about the thruster with the chosen semi-empirical development approach. Therefore, the development of the next Generation µHEMPT is based on the presented approach.

As first step a magnetic field topology was defined that summarises all the gained information and that could lead to a more efficient thruster. The chosen magnetic field is illustrated in figure 4.8. The illustration shows a half segment of the thruster discharge channel. The rotation axis is at the top, the thruster exit is at the right side. The magnet assembly is presented as grey structure. The flux density is given via the background colour plot. Bright blue illustrates zero flux density, whereas purple presents the highest flux density.

The cusp regions are tagged with the dashed circles. It can be seen that the field strength inside the cusp regions is maximised to improve the mirror possibility of the electrons which are bound to the magnetic field lines. Therefore, the mass utilisation of the thruster should be improved and the thruster should be able to operate with a minimum amount of propellant particles. To further increase the flux density, different pole shoe materials (such as pure iron, normalised carbon steel, or

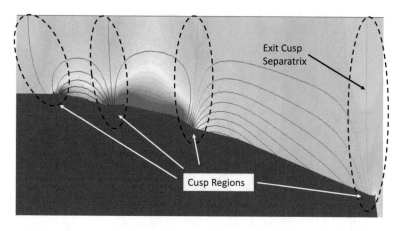

Figure 4.8: Illustration of the developed magnetic field geometry. The black lines present the magnetic field lines, the coloured background illustrates the flux density of the B-field, with bright blue equal to very low flux density and purple to high flux densities. The magnetic field is optimised for high electron confinement and low divergence losses.

mu-metal) and different magnet materials (SmCo or NdFeB) were considered. The maximum flux density at the middle cusp surface is 1.2 T

At the thruster exit, the separatrix of the exit cusp was designed to be as flat as possible. As mentioned, the flat separatrix leads to a high divergence efficiency of the thruster.

To support the qualitative assessment of the magnetic field topology, a particle tracing tool (SIMION) was used. The tool allows a simple tracing of single electrons inside B&E-fields. Figure 4.9 illustrates a representative result of the particle tracing projected inside the thruster discharge channel.

To track the particles with the particle tracer tool the B-field that was computed with FEMM, or Maxwell 3D was imported together with the CAD model of the thruster into the application. A static E-field that should replicate the E-field topology during thruster operation was also part of the simulation.

The simulation that is shown in figure 4.9 illustrates the traces of 5 different electrons. The electrons were started at the thruster exit with a velocity of 5 eV in a randomly assigned starting angle between ±90° with respect to the rotation axis of the thruster and in anode direction (to the right).

It can been seen how the electrons travel inside the B-field of the thruster and how they are reflected inside the cusps. If an electron enters the thruster it is accelerated by

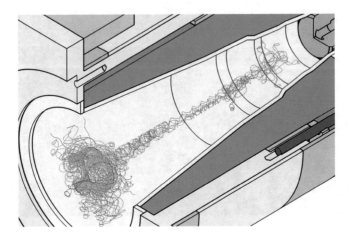

Figure 4.9: Result of a 3D electron tracing inside the thruster. The electrons are confined
inside the magnetic mirror structure which leads to low energy losses.

the E-field applied. This behaviour was also demonstrated in real plasma simulations
(particle in cell) of this thruster [71] and also with similar thrusters [111].

The high impedance between cathode and anode of the magnetic field tested,
which is a typical HEMPT feature, can be observed. The electrons which are started
at the outside of the thruster are mostly confined in the inside of the exit cusp.
This behaviour also illustrates the limits of the simple particle tracing simulation
compared to a real plasma simulation, because in reality the electron transport inside
the thruster is dominated by elastic collisions of the electron with neutral particles.
This also means that the real impedance of the thruster cannot be determined with
a simple particle tracing.

However, the simulation helps to achieve a qualitative overview about the behaviour of the electrons inside the thruster and is therefore a support to the semi-
empirical thruster development.

Relying upon the presented results, the development of a next generation µHEMPT
was started. Two different kinds of thruster have been built; a full size model that has
a minimal discharge chamber diameter of 12 mm and a maximal discharge chamber
diameter of 30 mm. The integrated thrusters are presented in figure 4.10. Figure (a)
shows the full scale model called NG-µHEMPT, whereas figure (b) illustrates the
four times smaller thruster called mini-NG-µHEMPT.

The larger engine should be used for a further studying of the HEMPT physics
without downscaling constraints. Moreover, the thruster should demonstrate the

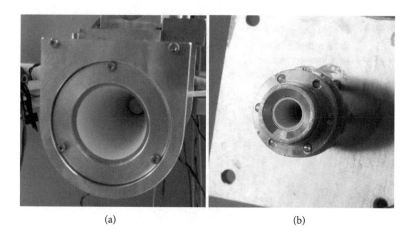

(a)   (b)

Figure 4.10: Pictures of the thruster models developed. In (a) the full scale model is shown. The ceramic discharge chamber and the anode made of brass at the end of the discharge chamber are visible. In picture (b) the miniaturised model is presented. At the inside of the thruster also the discharge chamber and the brass anode are visible.

quality of the presented magnetic field structure to again validate if the semi empirical thruster design approach works as expected for the NG-μHEMPT development.

Additionally, the miniaturised model that is four times linearly scaled down should demonstrate an operation in the micro-Newton regime. Therefore, it should demonstrate that it is possible to operate a HEMPT in the micro-Newton regime on a sufficient performance level. Furthermore, due to the two thruster models the effects of the linear downscaling can directly be observed and analysed.

4.2.1 Full Size NG-μHEMPT

Figure 4.11 provides an overview of the full size NG-μHEMPT developed. The magnetic field which was presented in figure 4.8 is formed by a magnet stack that consists mainly of SmCo magnets. Also NdFeB magnets have been tested in the beginning, however due to the high power dissipation of the thruster a measurable degradation of the temperature sensitive permanent magnets was induced. Thus, the full scale thruster data presented was exclusively generated with SmCo magnets.

The magnet stack is born by an aluminium structure so called the thruster structure. This structure holds the magnet stack in position in order to avoid asymmetries of the magnetic field which would lead to unpredictable changes of the electron

☐ Thruster Structure
■ Magnet Stack
■ Connection Rods
☐ Discharge Chamber
☐ Anode

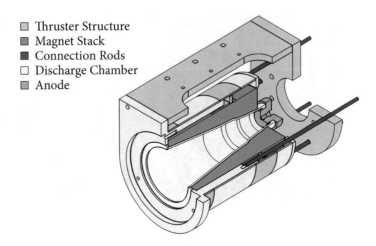

Figure 4.11: Illustration of the full size NG-µHEMPT developed. The slice out enables a view of the inside of the thruster.

confinement and therefore to unexpected variations of the thruster operation. The thruster structure also connects the engine to the test stand. It is optimised to be suitable for the testing inside the Airbus micro-Newton thruster test facility.

Additionally, the thruster can also be connected to other structures via the connection rods that are presented in blue.

The magnet stack and the thruster structure are isolated from the plasma inside the thruster via a ceramic discharge chamber. The chamber is illustrated in white. As opposed to the Alumina discharge chamber that was used in the EBB20, the new chamber was made of hexagonal boron nitride. It combines a good dielectric strength and a good secondary electron emission yield with an excellent processibility.

The good machining properties allow an adaptation of the ceramic chamber to the magnet stack and hence to the magnetic field lines. The conical discharge chamber has a minimal inner diameter of 12 mm at the anode and a maximal inner diameter of 30 mm at the thruster exit. It was also possible to reduce the wall thickness of the discharge chamber to 1 mm, i.e. the wall thickness to the minimal diameter ratio is 1/12 instead of 1/6 in case of the EBB20. Therefore, the quality of the magnetic mirror is improved and the electron wall losses inside the cusps are reduced.

According to the observations of the impact of the anode material on thruster performance of the EBB20 which were presented in [92], the anode used is made of brass. The anode is used as gas feed through in parallel to its function as anode of the plasma discharge. To feed the propellant into the chamber, four injector orifices are placed at the edges of the anode. A filter is placed at the backside of the anode to prevent plasma from back streaming.

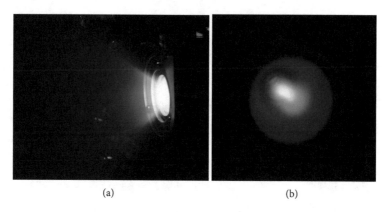

(a)   (b)

Figure 4.12: Pictures of the full scale NG- μHEMPT firing. In (a) the side view of the operating thruster is shown. In (b) the frontal view of the firing thruster is presented.

The thruster had been tested inside the Airbus micro-Newton thruster test facility. Due to the variable design of the facility it was possible to adapt all measurement instruments to the higher measurement regime that is required to operate the thruster. It was operated with mass flows up to 4 sccm.

As previously explained in chapter 3.3, the pumps of the facility were designed for a maximum gas ballast of 0.5 sccm, hence the background pressure during thruster operation was in the $1 \cdot 10^{-5}$ mbar regime. However, according to the general thruster testing practice of mN thrusters this background pressure should be sufficient for the tested thruster type [55].

The images that are presented in figure 4.12 illustrate the developed μHEMPT firing inside the test facility. The left picture (a) shows a side view of the thruster. Due to the atomic transition of the xenon used the plume shines in bright blue. The divergence of the plume is qualitatively contoured as brighter lines.

The picture on the right side (b) presents the view inside the discharge chamber of the thruster. The visible parts of the plasma are focused around the inside of the rotation axis. This behaviour is in agreement with the electron tracing performed and the published knowledge of the HEMPT physics [72, 112, 113]. The focusing of the plasma is caused by the electron confinement at the rotation axis. As a consequence a negative space charge would be established at the centre axis, but to maintain the quasi neutrality of the plasma, the ions follow the electrons to the middle axis. The confinement of the ions and the electrons lead to the almost erosionless operation of the HEMPT.

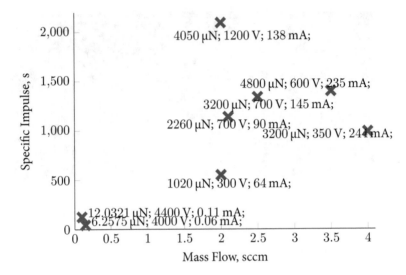

Figure 4.13: Summary of the tested operation points, where the specific impulse is plotted versus the mass flow. The thruster can be operated over a wide range. The operation point with the smallest achievable measured thrust is $6.2575\,\mu N$, with a low I_{sp} of only 65 s. At higher thrust levels I_{sp} of more than 2100, s can be demonstrated.

The thruster has been characterised to determine the operation space of the thruster and to identify the key figures of the thruster at different operation points. Figure 4.13 provides an overview of the observed thruster operation space where the specific impulse (in s) is plotted versus the mass flow (in sccm). Each operation point tested is marked with a red cross and beside the marked positions are the specific operation parameters presented. First the measured thrust is given followed by the applied anode voltage and the anode current.

In general, the NG-µHEMPT demonstrated a stable operation over a wide range. The propellant mass flow could be varied from 0.1 sccm to 4 sccm. Dependent on the chosen input parameters different thruster performances could be achieved with specific impulses of up to 2100 s. The typical PTTR of the thruster was around 30 W/mN.

Even at low mass flows (≤ 0.2 sccm) the plasma discharge was not terminated, hence also an operation with a small propellant flow was possible. However, the specific impulses in these operation modes are comparatively low. The best overall thruster performance was achieved at anode potentials around 1200 V. The mass utilisation is dependent on the input propellant flow and therefore different specific impulses can be achieved with similar anode potentials.

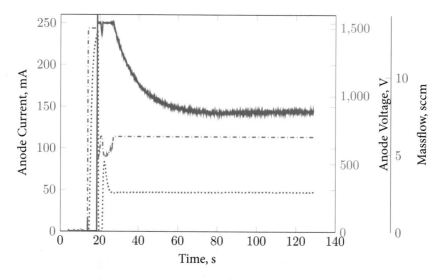

Figure 4.14: The plot presents an example of the thruster ignition. The anode current (solid line), the anode voltage (dash-dotted curve) and the mass flow (dotted line) are plotted versus time. The thruster is started by a short full opening of the mass flow controller.

The thruster demonstrated a good reproducibility of the used and tested operation points. To reach a specific working point, the thruster was throttled via the propellant mass flow.

It has been observed that a start-up of the thruster at higher anode potentials enables an uncomplicated thruster ignition. Hence, the thruster was typically started with higher anode potentials. Figure 4.14 illustrates the usual thruster start-up procedure. The plot consists of three curves: the anode current, the anode potential and the propellant mass flow are plotted versus time (in s). The anode current is shown as solid curve with the left y-axis. The anode potential is presented as dash-dotted line and has a separated y-axis at the right side. The measured propellant mass flow is illustrated as a dotted curve with a scale on the far right side.

To start the thruster, the anode potential was set to a high voltage level in the range of 1500 V. If the anode potential reached the defined maximum potential

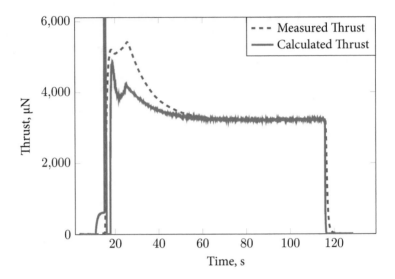

Figure 4.15: Representative thrust plot of a thrust measurement, where the force is plotted
against time. The measured thrust is shown in blue whereas the calculated
thrust is presented in green. After the ignition of the thruster, it has been
throttled to the targeted thrust. After 115 s the thruster was deactivated to
generate a new zero position that is used to validate the thrust measurement.

the valve of the mass flow controller was set to 90 % of its maximum aperture, thus
the measured mass flow rises rapidly. If the pressure inside the discharge chamber
reaches a critical limit the thruster ignites and runs into the current limit of the power
supply used, in the presented case 250 mA. If an ignition of the thruster is detected
the mass flow controller is set to closed loop operation, and the PID-controller that
is embedded in the laboratory mass flow controller starts to stabilise the mass flow
to the set value, in the presented case the set value is 2.7 sccm. With decreasing
mass flow the anode current also decreases. Thus, the power supply is set back to
the voltage limiting mode. In the presented figure the anode potential was set to
700 V after thruster start-up. With respect to the length of the propellant feed lines
used, which is usually 1.5 m in the Airbus test facility (flow controllers are placed
outside the vacuum chamber), the anode current normalises for almost 1 min. After
the successful ignition, the thruster can be throttled to each of the operation points.

After the throttling to the operation point that shall be tested, the thrust produced
can be measured with the thrust balance. An example of a thrust measurement is
given in figure 4.15 where the thrust (in μN) is plotted against time (in s).

The dashed curve illustrates the directly measured thrust. The solid curve represents the calculated thrust, calculated with the total efficiency and the input parameters of the thruster according to equation 2.3. The figure represents a full thrust measurement cycle, beginning with a deactivated thruster, the thruster ignition at $t = 15$ s, followed by the ignition procedure. After the normalisation of the thrust level the thruster was turned off to obtain a second zero inside the measurement data (at $t = 115$ s), which increases the confidence level of the measurement, although it would not be required due to the unique stability of the balance developed.

The curves illustrate the trend of the thrust produced during the start up phase. The ignition algorithm used produces an overshooting of the thrust. The thrust level requires 60 s to reach the commanded level.

Table 4.3: Summary of parameters that could be estimated via the direct thrust measurement at the presented operation point.

Parameter	Value
Anode voltage	700 V
Anode current	145 mA
Mass flow	2.54 sccm
Thrust	3236 μN
I_{sp}	1340 s
PTTR	31.20 W/mN
η_T	20.8 %

It can be observed that the measured thrust and the calculated thrust have qualitatively similar characteristics, but the absolute values are different before $t = 60$ s. This behaviour is caused by the thrust calculation method used that is based on a fixed total thruster efficiency, because typically the total efficiency is dependent on the specific operation point. Hence, in practice look up tables are used to translate the thruster parameters into a calculated thrust.

The trends of the presented curves suggest that at higher thrust levels the total efficiency of the thruster could be higher, since the measured thrust is higher than the calculated thrust.

Based on the direct thrust measurement, the thrust, the specific impulse, the PTTR and the total efficiency at the tested operation point can be determined. The result of a representative thruster operation point is summarised in table 4.3.

In addition to the direct thrust measurement the other loss term can be measured via the plasma diagnostic setup. The ion beam current and the divergence of the plume can be determined with the Faraday cup assembly. Figure 4.16 illustrates the

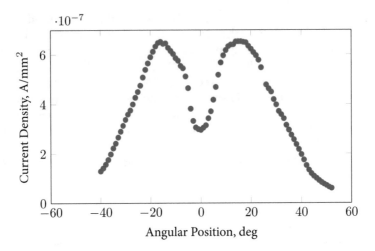

Figure 4.16: Result of the plume characterisation at 700 V anode potential and 145 mA anode current, where the current density is plotted versus the angular position.

result of the Faraday cup measurement at the previously discussed operation point. On the x-axis of the figure the angle is plotted, the current density is given on the y-axis. The rotation axis of the thruster is at $0\,°$.

The measurements were performed at a distance of 300 mm from the thruster. Each measurement point summarises a quasi static measurement of several seconds at the specific position. Thus, the given points are the mean value of 1000 data points. The measurement was performed between $50\,°$ and $-40\,°$. The distance between each measurement point is one degree.

The data set illustrates the shape of the plume at the specific operation point. According to the equations presented in section 2.4.2, the beam current, the discharge efficiency, the mass utilisation and the beam divergence can be calculated. The determined beam current is 123.25 mA and hence the discharge efficiency is 85 %. As mentioned, the input propellant mass flow is 2.55 sccm, thus the mass utilisation is 57.40 %. Assuming that 20 % of the ions that are generated inside the thruster are double charged ions, the divergence efficiency is 83.90 %.

In parallel to the Faraday cup measurement, the energy of the ions can be estimated via the RPA. A part of the measurement is presented in figure 4.17 where measured current density (in A/mm^2), at the collector of the RPA, is plotted against the applied retarding potential (in V). The four curves illustrate the gradient of the current density at different plume angles. The increasing retarding voltage the measured current density slowly decreases until the retarding potential reaches the ion potential

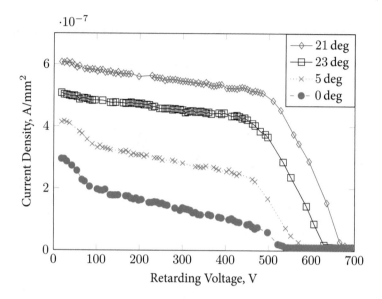

Figure 4.17: Exemplified RPA measurements at selected angular positions to illustrate the different ion energies at different positions. The Current density is plotted against the applied retarding voltage. The thruster was running at 700 V anode potential and 145 mA anode current.

and the current density drops to zero. Thereby, as mentioned in chapter 2.4.2, the acceleration efficiency can be determined via numerical differentiation.

The red curve presents the RPA measurement at 21 °. The maximum current density of the plume is at this angle, i.e. this position is of special importance for the overall thruster performance. Also the point where the current density drops to zero is at the highest potential in comparison to the other presented curves.

The data set that was obtained at 23 ° is illustrated in black. At this position the maximum current density is smaller than the current density in the red curve. This can also be seen in figure 4.16. The current density drop starts at a lower retarding potential, i.e. the acceleration efficiency is lower than the acceleration efficiency of the red curve.

The green and the blue curves present the measurement data at 5 ° and 0 °. It can be observed that the current density drops at two different potentials. Therefore, it can be assumed that on these angular positions, ions with different energy levels are measured by the RPA. The slow ions could be directly created in front of the thruster exit at plume space charge potentials or the slow ions could be generated by elastic or charge exchange collisions of the fast ions with slow neutral particles in front of the thruster.

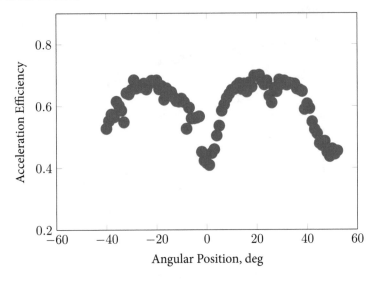

Figure 4.18: RPA measurements at various angular positions, the acceleration efficiency is plotted against the angular position.

The RPA measurements the applied retarding potential (in V) have been performed at each angular position to determine the acceleration efficiency distribution inside the ion beam. Figure 4.18 presents the result of the measurements. On the x-axis the angular position is plotted and the acceleration efficiency is plotted on the y-axis. The utilised accelerations for each angle are presented as dots. The maximum efficiencies are located between $30°$ and $10°$ and between $-10°$ and $-30°$. The distribution of the efficiencies is similar to the distribution of the current density of the ion beam.

Based on the ion current distribution, the total acceleration efficiency can be derived from the presented acceleration efficiency distribution and is 62.04 %.

The acceleration efficiency and the discharge efficiency can be combined to give the electrical efficiency of the thruster at the specific operation point, which is 52.73 %. Following to chapter 2, all information gained by the plasma diagnostic measurements can by transformed into a total efficiency, which is here 21.30 %.

The key figures that were derived from the direct thrust measurement and the indirect thrust measurement are summarised in table 4.4. The total efficiency that were determined via the indirect thrust measurement is almost equal to the total

Table 4.4: Example of the thruster parameters determined by the direct and indirect thrust measurement at one possible operation point. The small difference between η_T and $\eta_{T_{in}}$ underlines the quality of the measured data.

	Parameter	Value
	Anode voltage	700 V
	Anode current	145 mA
	Mass flow	2.54 sccm
	Thrust	3236 μN
Direct thrust	I_{sp}	1340 s
measurement	PTTR	31.20 W/mN
	η_T	20.8 %
	γ	83.90 %
	η_d	82.10 %
Indirect thrust	η_v	62.04 %
measurement	η_e	50.8 %
	η_m	57.88 %
	$\eta_{T_{in}}$	21.10 %

efficiency of the direct thrust measurement. This indicates that the measured data of the direct and indirect thrust measurement generate trustworthy results due to the fact that both measurements are independent from each other.

The key performance parameters of the thruster illustrate that the general thruster design including the selected magnetic field topology fulfils the targeted requirements such as a high specific impulse, a high divergence efficiency and a sufficient discharge efficiency. As expected the full scale thruster is not able to operate efficiently in the micro-Newton regime. However, an operation at these thrust levels is possible.

The best total efficiency which could be measured was 27.3 %, determined via a direct thrust measurement. The anode potential was 1200 V, the anode current was 135 mA and the propellant mass flow was 2 sccm.

In contrast to the EBB20, the acceleration efficiency measured is around 60 % and therefore 30 % lower than before. A reason for this reduction could be related to the expansion of the anode potential in case of the EBB20 downstream to the magnetic separatrix outside the thruster exit, whereas in the NG- μHEMPT a reasonable amount of ions are created at lower potential at the thruster exit separatrix. However, further analysis should be performed to fully understand the observed behaviour.

4.2.2 Miniaturised NG-µHEMPT

Since the very beginning of the HEMPT downscaling campaign it was clear that a linear downscaling from the milli-Newton to the micro-Newton thrust regime would decrease the overall efficiency of the thruster due to the fact that a linear geometrical scaling of the permanent magnets used would require an inversed linear scaling of the magnet magnetisation to maintain the Larmor radii ratios relative to the thruster dimensions. This is not feasible due to the currently available magnet materials. Id est, the Larmor radii of confined electrons are kept constant while the volume of the discharge chamber is a significantly reduced. Hence, higher electron losses occur. This effect is amplified because of the manufacturing limitations mentioned whereby the ceramic discharge chamber does not allow a simple linear scaling of ceramic wall thickness in parallel to the magnet stack down-scaling.

However, a reduction of the discharge chamber diameter is required to maintain the neutral gas density in the inside of the thruster that leads to a sufficient mass utilisation and thus to a reasonable specific impulse. In order to overcome this limitation a miniaturised NG-µHEMPT has been built and characterised.

Figure 4.19 presents the mechanical design of the laboratory model developed. The magnet stack (dark grey) is mounted via an aluminium structure (light grey). As mechanical interface threaded rods are used which are illustrated in blue. The rods are made of brass to avoid interactions of the rods with the magnetic field. At the end of the discharge chamber (shown in white), the anode (brass coloured) is placed. Like in the EBB20 and the full scale model, the anode is used in parallel as the gas inlet.

The maximum diameter of the discharge chamber (without ceramic wall) at the exit magnet is 7.5 mm. The minimum diameter directly at the anode is 3.5 mm. The overall length of the discharge chamber is 16 mm. Thus, the diameter of the mini NG-µHEMPT has the same scale as the EBB20 which should lead to an operation with minimal propellant feeding and thus a minimal thrust.

The numerical magnetic field simulations were also performed for the four times smaller magnet stack. The result of the simulation and especially the shape of the magnetic field lines have not been changed in comparison to the full scale thruster, i.e. the magnetic field topology which is presented in figure 4.8 represents the magnetic field topology of the miniaturised µHEMPT.

The improved magnetic field topology should improve the divergence efficiency and reduces the electron losses in comparison with the EBB20 and the EBB20 MkII. Additionally, a minimal wall thickness of the ceramic discharge chamber is required to reduce the wall loss and to improve the electron confinement inside the thruster. As mentioned, the minimal wall thickness is dependent on the chamber material used and the tolerances of the mechanical machining processes. In house

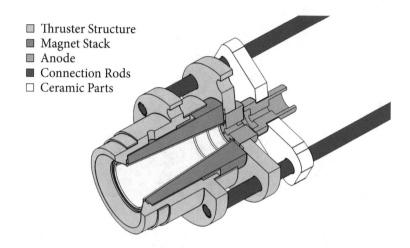

- ▢ Thruster Structure
- ◼ Magnet Stack
- ▢ Anode
- ◼ Connection Rods
- ☐ Ceramic Parts

Figure 4.19: Illustration of the miniaturised NG-µHEMPT; the model is four times smaller than the full scale model. The thruster is designed to maintain the magnetic field geometry and thus the magnet stack is scaled down linearly. The mechanical structure of the thruster could not be scaled down due to the manufacturing and assembling of the thruster.

investigations have demonstrated that a ceramic wall thickness in the order of some tenth of millimetres is extremely hard to manufacture.

To overcome this limitation, the magnets were only coated by a thin Boron-Nitride layer. The method developed allows the deposition of single or multiple non conducting Boron-Nitride layers with a thickness of $< 10\,\mu\text{m}$. Figure 4.20 illustrates the effect of the ceramic coating of the discharge chamber; in (a) an example is presented with a conventional solid ceramic discharge chamber, whereas (b) presents the coated discharge chamber dimensions.

As mentioned, at smaller thruster dimensions the manufacturing tolerances of the solid discharge chamber lead to a higher wall thickness to thruster diameter ratio. The ceramic coating solves this problem. Hence, the electron losses due to wall collisions are minimised which leads to an operation with a reduction of the propellant mass flow compared to the previously developed µHEMPTs.

For a further improvement of the electron confinement NdFeB-magnets have been used instead of SmCo-magnets for the magnet stack. As mentioned, the Curie temperature of the NdFeB-magnets is lower than the Curie temperature of the SmCo magnets. Thus, the input power to the thruster has to be as small as possible to

■ Thruster Structure ■ Discharge Chamber

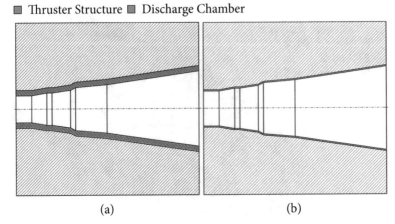

(a)   (b)

Figure 4.20: Illustration between a conventionally manufactured discharge chamber (left side) and the coated discharge chamber. The coating enables a minimal wall thickness. Hence, the electron losses inside the discharge chamber are reduced.

reduce reversible and irreversible reduction of the flux density which would cancel the positive effects of the NdFeB magnets (see figure 4.7)

The developed thruster was integrated and tested at the Airbus micro-Newton thruster test facility to characterise the thruster to estimate the performance of the thruster and to validate the assumptions of the semi empirical development approach used. The thruster firing is shown in figure 4.21, photography presents the side view (left) and the front view. The plasma discharge has the same shape than the discharge of the full scale thruster (see figure 4.12) with a strong confinement of the visible plasma in the centre axis.

The thruster was operated with a COTS laboratory power supply (FUG HCP 140-20000) which has a maximum output voltage of 20 kV. The power supply can be used in voltage limited mode and in current limited mode. The power supply offers a small voltage ripple of $1 \cdot 10^{-5}$ V [114] and a maximum output current of 6 mA. The relatively small output current avoids an overheating of the thruster i.e. it protects the magnet stack from an irreversible damage.

To control the propellant feeding a COTS mass flow controller (Bronkhorst El-flow F-200CV) was used. The controller has a minimal output mass flow of 0.02 sccm and a maximum output of 1 sccm, calibrated on xenon. The resolution of the controller is 1:50 of the maximum output flow rate; for example the relative error at an output mass flow of 0.1 sccm is 20 %.

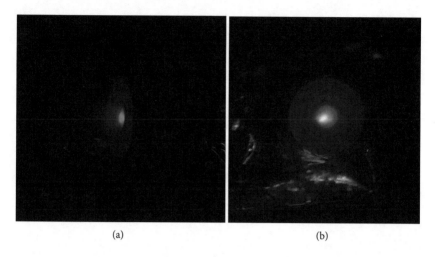

(a)   (b)

Figure 4.21: Pictures of the firing miniaturised NG- μHEMPT. In (a) the side view of the
operating thruster is shown. In (b) the frontal view of the firing thruster is
presented.

With respect to the goals of the miniaturised NG-μHEMPT development, the
thruster was especially tested with small mass flows, although an operation at higher
propellant feeding levels would also be possible. However, to keep the thermal load
on the thruster as small as possible, these operation points were not characterised.

In figure 4.22 the operation points tested are presented. On the y-axis the specific
impulse is plotted against the mass flow. Each tested point is marked by a red cross.
Next to the crosses, the operation parameters measured are written. The first value
is the direct measured absolute thrust, the second value is the anode potential and
the third parameter is the anode current. The direct measured thrust has been used
to calculate the specific impulse and the total efficiency of the thruster; which is in
the range of 3 % to 7 %.

The minimal mass flow that had been used for detailed testing was 0.1 sccm,
because until this mass flow level the thruster presented a stable operation. Hence,
detailed measurements could be performed. Directly at 0.1 sccm, the thruster could
be operated in a range between 29 μN and 86 μN. The thruster could be mainly
throttled by the variation of the anode voltage. At higher anode voltages also the
anode currents increased.

It has been observed that the position and relative emission current of the electron
sources used (mainly for beam neutralisation) also influenced the thruster operation.
As electron source, simple tungsten filaments as hot cathodes were used. In particular,

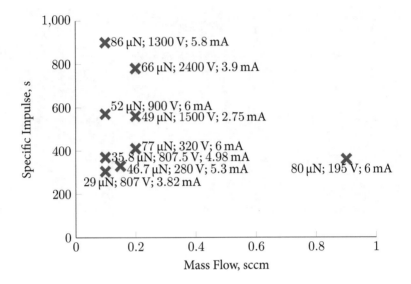

Figure 4.22: Summary of the tested operation points, where the specific impulse is plotted
versus the mass flow. The thruster can be operated over a wide range. The
operation point with the smallest achievable measured thrust is 29 µN, with a
low I_{sp} of only 295 s. At higher thrust levels Isps of more than 900 s could be
demonstrated.

at the low thrust operation points, the coupling between electron source and thruster
was remarkable. The points at 29 µN and 35.8 µN at almost the same anode potential
illustrate this relationship.

The operation point with the highest specific impulse was measured at a flow
rate of 0.1 sccm at 86 µN with an anode voltage of 1300 V and an anode current of
5.8 mA. The total efficiency determined was 5.5 %. These parameters are summarised in table 4.5.

Different tests were also performed at 0.2 sccm propellant mass flow. . It was
possible to operate the thruster at different anode potentials and anode current
levels. At higher mass flows the thruster could be operated, but due to the limited
maximum anode current the overall thruster performance is constraint at higher
propellant feeding rates.

To illustrate the start up characteristic of the miniaturised thruster developed,
figure 4.23 provides an overview of the basic thruster parameters. Similar to the
full scale engine the ignition is triggered by a short full opening of the mass flow
controller valve used. The measured mass flow rate is shown as the dotted curve

Table 4.5: Illustration of the parameters of one single operation point estimated via the direct thrust measurement. These parameters were measured for all operation points presented.

Parameter	Value
Anode voltage	1300 V
Anode current	5.8 mA
Mass flow	0.1 sccm
Thrust	86 μN
I_{sp}	897 s
PTTR	87.7 W/mN
η_T	5.5 %

(red) with the far right scale (red). In parallel to the short mass flow peak, the anode potential is set to 1800 V. The measured anode potential is presented as dash-dotted curve (green) and the scale on the right side.

Due to the long propellant feeding line between the mass flow controller and the thruster in the laboratory setup (1.5 m), the delay between thruster ignition and full opening of the propellant valve is in the order of several seconds. After the ignition of the thruster the mass flow controller set point can be directly reduced, in figure 4.23 the flow rate was set to 0.3 sccm.

The ignition of the thruster can be recognised in figure 4.23 at 49 s were the anode current jumps from zero to the maximum value of 6 mA. The thruster operates in current limiting mode, i.e. the anode potential is set self-consistent. With respect of the length of the propellant feeding line the dead time between measured mass flow and real mass flow at the thruster is in the order of several seconds. Thus the reaction of the thruster on mass flow variations is delayed. In order to take this into account, a waiting time of 20 seconds is inserted into the ignition control loop. After this time the mass flow is throttled down stepwise to the targeted mass flow. These steps are also performed with a 20 seconds delay.

Simultaneously, the thruster can be throttled to the targeted operation point via the anode voltage and the anode current. The presented operation point has an anode potential of 1500 V, an anode current of 2.7 mA and a mass flow of 0.1 sccm.

The whole start-up procedure requires 140 s to reach the targeted operation point after thruster ignition.

The operation point described illustrates the ambiguous behaviour of the anode voltage in comparison to other points that are presented in figure 4.22 i.e. a higher anode voltage does not automatically lead to higher anode currents. Depending on

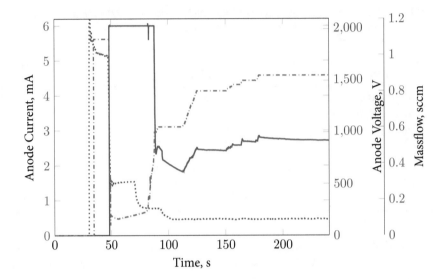

Figure 4.23: An example of the ignition of miniaturised NG-µHEMPT. The anode current
(solid, blue), the anode voltage (dashed-dotted, green) and the mass flow
(dotted, red) are plotted against time. The ignition is triggered by a short full
opening of the mass flow controller.

the available neutraliser electrons and the propellant mass flow, a high anode voltage
can also lead to a lower anode current, that is typically equal to a lower discharge
current and therefore to a reduced mass efficiency.

As mentioned, for all presented operation points a direct thrust measurement has
been performed. Figure 4.24 presents a representative thrust measurement example
which was recorded simultaneously to the curves of figure 4.23.

The plot presents the thrust (in µN) versus time (in s). The blue curve presents
the measured absolute thrust, whereas the green curve presents the calculated thrust
derived from the anode potential, the anode current and the propellant mass flow.

After five seconds the propellant mass flow valve was opened (after thirty seconds
in figure 4.23). The thrust balance measures the thrust of the neutral gas flow which
leads to a first small thrust measurement signal after ten seconds. The thruster ignition leads to a fast rising of the calculated thrust, whereas the measured thrust rises
more slowly with respect to the transfer function of the thrust balance, but also of the
transfer function of the thruster and the calculation method used (mainly influenced
by the long propellant feeding line) which is not considered in the calculated thrust
curve.

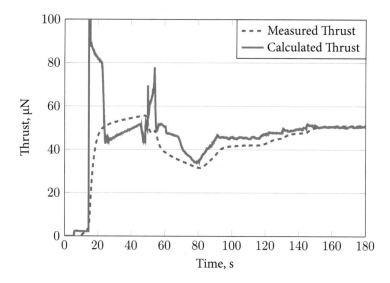

Figure 4.24: Representative plot of a thrust measurement of the miniaturised thruster, where the force is plotted against time. The measured thrust is shown in blue whereas the calculated thrust is presented in green. After the ignition of the thruster, it has been throttled via different operation points to the targeted thrust level. After 160 s the thruster reached the targeted thrust level.

The start-up procedure of the thruster requires 140 s to reach a stable operation (see also figure 4.23). The measured absolute thrust level at this point is used to calculate the total efficiency of the thruster at the targeted operation point. In this example the total efficiency is 3.1 %.

The differences of the measured thrust and the calculated thrust resulted from variation of the total efficiency of the thruster; each operation point has a different total efficiency. Moreover, as mentioned, the transfer function of the thruster (especially the transfer function of the propellant feeding) is not taken into account. Thus, especially at the ignition, much higher thrust levels are calculated due to the high measured gas flow rate. However, these calculated values are considered as invalid data because the real propellant mass flow rate at the anode cannot be measured which leads to high uncertainties.

In parallel to the direct thrust measurements a characterisation of the thruster with the plasma diagnostic setup was also performed. A mapping of the plume is presented in figure 4.25 where the normalised current density is plotted against the planar angle. The current density of each curve was normalised to the maximum

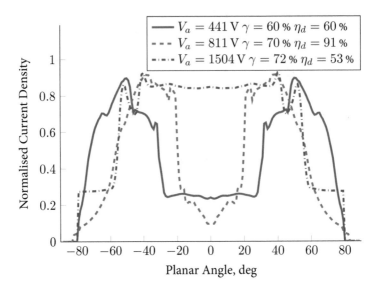

Figure 4.25: The plume shapes of the mini-NG-µHEMPT at different anode voltage levels at a constant propellant mass flow of 0.1 sccm. The current density normalised is plotted versus the planar angle position in degrees. The curves illustrate that the shape of the plume and therefore the divergence efficiency is strongly dependent on the anode voltage.

measured current density which was measured in each data set. All measurements have been performed at a constant propellant mass flow of 0.1 sccm. In all curves the measured beam current density at angular positions larger 80 degree is zero or almost zero. This is the result of the thruster structure which blocks the ion beam at angles wider than 80 degree. The figure enables a comparison of the plume shape at different anode potential levels.

As solid (blue) curve the measured angular current density distribution at 441 V anode potential and 6 mA anode current is presented. The processing of the measured data leads to a divergence efficiency of $60\,\%$ and a discharge efficiency of $60\,\%$. In the centre of the plume a current density gap can be observed. The plume has a shape which is qualitatively comparable to the plume of the full size NG-µHEMPT model (see figure 4.16) but the ion beam has a wider divergence angle.

The measurement result at 811 V anode potential and 5.4 mA anode current is given as the dashed (green) curve. It can be observed that the plume is more focused to the rotation axis which results in an increased divergence efficiency of $70\,\%$. Moreover, the discharge efficiency is $91\,\%$, i.e. it is also increased.

The dashed-dotted (red) curve in figure 4.25 illustrates the shape of the plume at an anode potential of 1504 V and an anode current of 2.75 mA. The current density in the centre of the plume is increased compared to the lower potentials. The gap that can be observed at 441 V and 811 V is not visible any more. Therefore, the divergence efficiency is slightly higher than on lower anode potential levels, it is 72 %. Although the beam has a more Gaussian shape, the increase of the divergence efficiency is small in comparison to lower anode potentials because of the qualitatively high amount of current density at higher angular positions. In contrast to the increased divergence efficiency, the discharge efficiency is reduced to 53 %. Therefore, the total efficiency at this operation point is slightly lower than the 811 V operation point.

Figure 4.25 shows that although the magnet separatrix has not changed in comparison to the full scale NG-μHEMPT the divergence of the beam is higher. This can be explained by the fact that due to the linear downscaling of the thruster size without increasing of the absolute magnetic field strength the impact of the magnetic electron confinement is reduced. As mentioned, this is a result of the changed Larmor radii to thruster dimension ratio. As a result the influence of the separatrix shape on the beam divergence is reduced which leads to lower divergence efficiencies.

Since the impact of the magnetic field is small, it must be assumed that the plasma potential characteristic has changed in comparison to the full scale NG-μHEMPT. It is possible that the potential forms a protuberance at the thruster exit, similar to the EBB20, that encourages the larger beam divergence.

Due to the protuberance of the plasma potential the observed strong dependence of the anode potential on the beam divergence angle also could be explained. Higher potentials result in an increased charging of the ceramic discharge chamber wall. This positive wall potential could than lead to an electrostatic focusing of the ion beam.

Currently not fully understood is the decreased discharge efficiency at higher anode potentials. Possible reasons could be an electron loss via a leakage current over the coated discharge chamber walls, or a higher loss of electrons due to the higher anode potential.

In addition to the Faraday cup measurements a characterisation of the acceleration efficiency was also performed. The result of a representative measurement at an anode potential of 441 V and an anode current of 6 mA is presented in figure 4.26. The plot shows the acceleration efficiency versus the planar angle. The overall acceleration efficiency is in the range of 70 %.

At angular positions greater than 80 degrees the current density measured, as presented in figure 4.25, was too small to perform a RPA measurement, thus no data can be provided for angular positions greater than 80 degrees.

In the centre of the ion beam where the current density is reduced the acceleration efficiency is also reduced. At the angular positions with higher current densities the

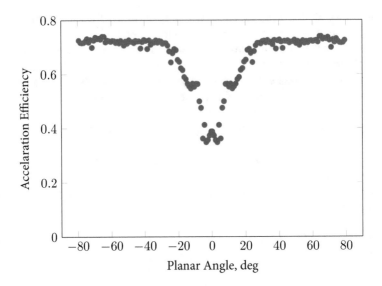

Figure 4.26: The figure summarises the results of one performed RPA measurements of the developed mini NG-µHEMPT at various angular positions, the acceleration efficiency is plotted versus the angular position.

acceleration efficiencies are increased. Measurements at different operation points have shown that the overall acceleration efficiency remains at a comparable level. The relation between the ion current density and the acceleration efficiency is also similar at different operation points.

It can be observed that the acceleration efficiency remains high also at the outer regions of the plume. This was not observed on the full scale NG-µHEMPT. This measurement result could also be explained with the protuberance of the plasma potential at the thruster exit. Thus, also at greater angular positions ions are able pass a high potential gradient. Therefore, the measured shape of the angular distribution is caused by the reduced influence of the magnetic electron confinement and hence a direct result of the linear downscaling.

The efficiency of the miniaturised thruster at the tested operation points can be calculated. The values of one of the tested operation points are summarised in table 4.6. Additionally to the indirect thrust measurement, a direct thrust measurement has also been performed to validate the results.

It can be seen that the total efficiency of the direct thrust measurement is with 5.10 % comparable to the estimated total efficiency of the indirect thrust measurement which is 6.67 %. This value is in between the predicted uncertainty budget of

Table 4.6: Summary and comparison of the thruster parameters determined by the direct and indirect thrust measurement for one example operation point. The difference between η_T and $\eta_{T_{in}}$ is 1.67 %

	Parameter	Value
	Anode voltage	441 V
	Anode current	5.8 mA
	Mass flow	0.1 sccm
Direct thrust measurement	Thrust	51 µN
	I_{sp}	532 s
	PTTR	51 W/mN
	η_T	5.1 %
Indirect thrust measurement	γ	60 %
	η_d	59.6 %
	η_v	68 %
	η_e	44.3 %
	η_m	47.7 %
	$\eta_{T_{in}}$	6.67 %

the thruster balance and the plasma diagnostic tools. The data presented is validated due to the good agreement between both measurement techniques.

In addition, to the absolute direct thrust measurements performed, thrust noise measurements were performed with the miniaturised engine, whereby, the thruster was operated continuously for more than 20 h without any restart.

The result of the measurement performed is presented in figure 4.27 were the normalised thrust noise is plotted against the frequency. The PSDs shown are derived from a single 20 h measurement which was down-sampled to 20 Hz. To generate the plot a Blackman-Harris 92 window has been used [10].

The measurement was performed at an anode potential of 670 V with an anode current of 4.2 mA. To control the thrust, the mass flow was used. Thus, a digital PI-controller has been implemented to stabilise the anode current at the mentioned 4.2 mA. Due to the long tube distance between the mass flow controller and the thruster the transient time of the control loop was long. Hence, the controller was not able to control short term anode current fluctuations. For a correct interpretation of the curves that are presented, it is important to mention that the direct thrust noise measurement summarises all potential noise sources within the thruster system, such as the power supply used, the mass flow controller including pipe work and filters, the thruster controller and the thruster itself.

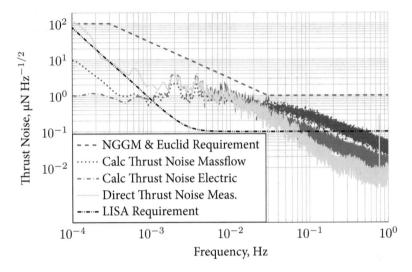

Figure 4.27: The curve presents a direct thrust noise measurement (solid plot) of the miniaturised NG-µHEMPT in comparison with the targeted requirements (dash-dot-dotted and dashed curves) and two different indirect estimated noise figures (dotted and dash-dotted).

In figure 4.27, the dashed (blue) curve illustrates the thrust noise requirement of NGGM and Euclid. The dash-dot-dotted (black) curve presents the thrust noise requirement of LISA. Both requirements were introduced in chapter 1. The PSD that is shown as the solid (yellow) curve presents the result of the direct thrust noise measurement of the miniaturised NG-µHEMPT system.

Between $1\,\text{Hz}$ and $81 \cdot 10^{-2}\,\text{Hz}$ the measured thrust noise is below $0.1\,\mu\text{N}$. At lower frequencies the thrust noise increases to $1\,\mu\text{N}$. At frequencies lower than $1 \cdot 10^{-3}\,\text{Hz}$ pink noise starts to dominate the measured thrust noise.

Hence, the tested miniaturised µHEMPT fulfils the LISA requirement at frequencies between $1\,\text{Hz}$ and $8 \cdot 10^{-2}\,\text{Hz}$. Below this frequencies the thruster is not compliant with this requirement. However, the thruster system fulfils the thrust noise requirement of NGGM and Euclid over the whole measurement bandwidth.

To compare the directly measured thrust noise performance with the indirect measured thrust noise the thrust noise that has been derived from the parallel measured mass flow, anode potential and anode current are presented in figure 4.27 as the dotted (red) curve. The thrust has been calculated with equation 2.3. The indirect measured thrust noise is also called calculated thrust noise. It can been seen that between $1\,\text{Hz}$ and $3 \cdot 10^{-2}\,\text{Hz}$ the calculated thrust noise level is higher than the directly measured thrust noise level. Between $3 \cdot 10^{-2}$ Hz to $1 \cdot 10^{-3}$ Hz the thrust noise level is equal. After $1 \cdot 10^{-3}$ Hz the calculated thrust noise level remains constant whereas the directly measured thrust noise level starts to rise.

One reason for the higher calculated thrust noise level is the long propellant feeding line that acts as a low pass filter which is not considered within the presented curve. The measured noise of the mass flow must be higher than the real noise of the mass flow at the thruster. The curves illustrate that this difference falsifies the indirect thrust noise measurement. In order to avoid this, a thrust noise estimation which is exclusively based on the measured anode potential and the anode current has been performed as well. It is presented as the dash-dotted (green) curve. As expected the thrust noise level is reduced in comparison to the dotted curve. However, the thrust noise level between 1 Hz and $3 \cdot 10^{-2}$ Hz is still higher than the directly measured thrust noise (solid curve). Thus, the measured anode current noise must be higher than the real noise of the ion beam.

The results underline that the thrust noise calculation via the simple assessment of the anode potential and the anode current is not sufficient to estimate the real thrust noise performance for the tested thruster. Depending on the measurement principle and the measurement positions of the anode current, anode potential and propellant mass flow the indirect thrust noise measurement could generate invalid measurement result; whereas the direct thrust noise measurement is able to characterise the complete thruster system.

The measurements also illustrate that all components of the thruster system influence the thrust noise performance of the thruster system. Thus, the measurements do not finally allow a statement of whether or not the miniaturised NG-µHEMPT can be used as a thruster for highly precise AOCS because the laboratory hardware used and especially the thruster controller, was not in a sufficient development stage to judge the results.

4.3 NG-µHEMPT Development Summary

In the previous sections two different NG-µHEMPTs have been presented and the key operation and performances figures have been analysed. It has been shown that due to the geometrical scaling different operation regimes could be accomplished. In figure 4.28 a summary of the combined operation space of the full scale and miniaturised thruster is presented. The plot shows the specific impulse versus the mass flow. Each cross illustrates a characterised operation point where a direct thrust measurement and a plume characterisation have been performed.

Figure 4.28 underlines that the thrusters developed are able to operate over a wide operation range and that the targeted operation space was achieved.

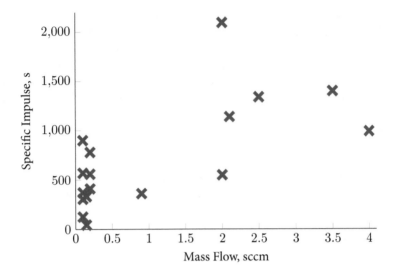

Figure 4.28: Illustration of the operation space which has been achieved with the developed NG-µHEMPTs. The figure sums up the figures 4.22 and 4.13.

Table 4.7 provides an overview of the operation parameters of both thruster variants. The full scale µHEMPT demonstrated a stable operation over a wide range. The anode potential can be varied from very low potentials (150 V) to relatively high anode potentials (4400 V) i.e. the thruster can be operated with different specific impulses and PTTRs because these figures are closely coupled with the anode potential.

The range of the anode potential of the four times smaller thruster is not as wide as the full scale thruster. However, with a range from 195 V to 2400 V the small thruster can be operated at various anode potentials which enables a reasonable throttle range and the operation with different specific impulses and PTTRs.

As mentioned, especially the full scale NG-µHEMPT showed an extremely stable operation during testing, this fact is underlined by the anode currents and the mass flow levels on which the thruster was operated (see table 4.7). The wide range of the usable anode potential, anode current and propellant mass flow leads to a large throttability of the engine. As also presented in table 4.7 the thruster can be operated over almost three orders of magnitudes of thrust.

The maximum thrust level of the full scale laboratory thruster could be extended with an improved thermal design of the thruster by radiative cooling or water cooling (only for laboratory testing) of the magnet stack. Thus, the maximum input power

Table 4.7: Comparison of the operation parameters of the NG-μHEMPT variants developed.

Type	Anode potential range		Anode current range		Mass flow		Thrust range	
Full scale	150 V 4400 V	-	0.06 mA 250 mA	-	0.1 sccm 4 sccm	-	6.275 μN 4800 μN	-
Mini	195 V 2400 V	-	1 mA - 6 mA		0.1 sccm 0.9 sccm	-	29 μN - 86 μN	

could be increased and therefore the maximum thrust level. It was also observed that the best total efficiencies were achieved at higher thrust levels, i.e a thruster operation at even higher thrust levels could also lead to a higher total efficiency of the thruster.

In comparison, the miniaturised thruster was not able to be operated over several orders of magnitudes of thrust. The thrust was varied between 29 μN and 86 μN with a maximum anode current of 6 mA. In general, it should be possible to operate the thruster on higher anode current levels, but the power supply that was used for thruster testing had a maximum anode current of 6 mA. As mentioned, the power supply was selected due to the targeted thrust regime and to limit the thermal load of the thrusters NdFeB magnet stack. An optimisation of thermal design of the miniaturised thruster should be performed to increase the operation space, like suggested for the full scale model.

The key performance figures of the thruster developed are presented in table 4.8. The table allows a comparison of the full scale thruster and the miniaturised thruster. It is clear that the almost linear downscaling of the thruster leads to a reduction of all key figures assessed. However, the difference between the values presented are not scaling with the factor of the downscaling, except of the total efficiency (η_T) which is more than four times lower. The best achievable specific impulse scales with a factor 2.3, the PTTR scales with a factor two and the thrust level at the operation point with the highest efficiency scales with a factor of 20. Especially the miniaturised thruster should be able to operate at higher thrust levels and a higher anode current which could lead to operation points that are potentially more efficient than the currently characterised operation points.

The values illustrate the performance degradation which is induced by the higher losses due to the downscaling. These effects have already been observed in A. Keller's downscaling study [32]. But the impact of the linear downscaling see in this development was not as large as in the study mentioned. Therefore, it can be assumed that the design adoptions such as the Boron-Nitride coating and the use of NdFeB

Table 4.8: Key figures of the developed NG-µHEMPT variations. The almost linear geometrical downscaling of the thruster leads to a reduction of all performance figures which can be explained with higher wall loss.

Type	Best I_{sp}	Typ. PTTR	Opt. Thrust	Typ. η_T	Typ. γ
Full Scale	2100 s	32 W/mN	4000 µN	25 %	85 %
Mini	900 s	65 W/mN	86 µN	6 %	70 %

magnets limit the downscaling effects to a sustainable level, although they cannot totally compensate the increasing loss processes.

It can be assumed that the electron losses inside the thruster are the dominating loss process, due to the observed operation space and the hands-on experience which was gained during thruster operation. This assumption is based on the fact that the operation range of the miniaturised thruster, especially at very low thrust operation, is smaller than the operation range of the full scale thruster. The lower electron losses lead to a more stable thruster operation, especially at very low neutral gas pressures inside the discharge chamber.

The data also illustrates that the reduced electron confinement inside the downscaled magnetic field leads to a reduced overall impact of the magnetic field. This leads to higher divergence losses of the miniaturised thruster even though the magnetic field line topology is preserved. Additionally, the smaller influence of the magnetic field could potentially lead to a changed plasma potential characteristic of the thruster with a protuberance of the thruster exit that could also explain the lower divergence efficiency and also that at larger angular positions ions with a high energy were measured.

However, the presented data do not allow a detailed theoretical understanding of what the real efficiency limiting process is, because not all necessary plasma parameters were measured. Detailed numerical simulations could also be used to enhance the theoretical understanding. Such simulations could be validated with the assessed experimental data. This kind of simulations are currently ongoing at the University of Bremen [72].

The miniaturised thruster demonstrated that an operation in the micro-Newton thrust regime with a sufficient thruster efficiency is possible. To illustrate this statement, table 4.9 summarises the targeted thruster development key figures with the measured data. The thruster is able to generate less than 50 µN, thus the thruster fulfils this development goal. With 900 s specific impulse the thruster almost fulfils the targeted requirement. Although the full scale thruster shows a sufficient divergence efficiency, the miniaturised thruster does not fulfil the targeted requirement, but there is an improvement compared with the divergence efficiency of the previously developed μHEMPTs.

The high PTTR leads to a total efficiency which does not fulfil the targeted efficiency. However, compared with other thruster technologies which rely on comparable physics (e.g. Hall Effect thrusters) the total efficiency achieved could be sufficient for the targeted applications, especially because of the extremely simple thruster system design.

As already mentioned, the thrust noise performance measured only fulfils the LISA thrust noise requirements in a small bandwidth between 1 Hz - 0.1 Hz. However, the result was measured with a simple anode current controller that could be the reason for the non compliant thrust noise performance of the engine. It can be assumed that a system that is optimised for low thrust noise operation, for example with a much smaller propellant feeding tube line and a real system controller, should be able to fulfil the requirement in the whole measurement bandwidth.

Table 4.9: The NG-μHEMPT performance achieved compared with the goals of the thruster development.

Parameter	Targeted value	Achieved value
Minimum thrust	$\leq 50\,\mu N$	$\leq 50\,\mu N$
I_{sp} in low thrust	$\geq 1000\,s$	$\geq 900\,s$
PTTR	$30\,W/mN$	$60\,W/mN$
Divergence efficiency	$\geq 85\,\%$	$\geq 70\,\%$
Total efficiency	$\geq 20\,\%$	$\geq 6\,\%$
Thrust noise	$\leq 0.1\,\mu N/\sqrt{Hz}$	$\leq 0.1\,\mu N/\sqrt{Hz}$ (1 Hz - 0.1 Hz)

The demonstrated performance figures of both thrusters underline that they are able to operate efficiently in the targeted thrust regimes.

The full scale NG-μHEMPT presents a promising performance in the milli-Newton range. The thruster is 1.5 - 2 times more efficient than a Hall Effect thruster when it operates at the same operation point [115, 116]. In the thrust envelope presented a high specific impulse paired with a comparatively high total efficiency could be realised. Additionally, an operation with a lower specific impulse and an improved PTTR was demonstrated. Hence, a further development of the laboratory thruster model to a single, or to more specific operation points seems to be feasible.

Simultaneously, the miniaturised NG-µHEMPT demonstrated that it is possible to use a thruster which relies on the HEMPT concept to produce ultra low thrust. The demonstrated I_{sp} of 900 s is of course not comparable with the I_{sp} of FEEPs or RITs. But, by taking into account that the system complexity of the thruster is almost as low as a cold gas thruster, the miniaturised laboratory model works with 18 times higher propellant efficiency than the nitrogen cold gas system that is used on the LISA pathfinder mission [43, 44, 9]. Compared with a cold gas thruster system that uses xenon as propellant the developed thruster is 30 times more propellant efficient.

As mentioned, the overall system complexity and mass of the developed miniaturised µHEMPT is extremely low and therefore comparable to a cold gas thruster system. The use of miniaturised µHEMPT could be an attractive alternative to other EP technologies which have a higher system mass and complexity. The predicted almost unlimited life time of the thruster is another advantage which has to be considered. The laboratory model could also be used to develop a cold gas thruster that is able to operate in a boosted mode, for mission scenarios which require very low thrust with a very low specific impulse and higher thrust levels paired with a higher specific impulse. The low specific impulse in very low thrust operation could be tolerated due the overall system mass (thruster system mass plus propellant mass).

From the performance figures presented for the µHEMPTs developed, a further development of the µHEMPT seems to be promising. Both thruster types, milli-Newton and micro-Newton, are predestined for different use cases, e.g. the full scale NG-µHEMPT could be used for small satellites in the 100 kg to 200 kg class whereas the miniaturised NG-µHEMPT could be used for CubeSats (e.g. used with Iodine as propellant) and small, or medium science and Earth observation missions.

5 Conclusion and Outlook

In this thesis the development, design and characterisation of a micro-Newton thruster test facility was presented. The facility consists of the required vacuum infrastructure, a highly precise and highly stable micro-Newton thrust balance and a set of plasma diagnostics. Measurement results of real thruster tests that has been performed with the facility has been shown and analysed.

As a second part the development of the next generation µHEMPT was also part of this thesis. Two next generation thrusters has been developed and tested in parallel, a full scale thruster and a miniaturised thruster. Before this development was started, as a precursor, an enhanced thruster based on the thruster design by A. Keller has been built and tested as well. This thruster was used to validate the chosen semi empirical development approach which was used to develop the next generation thrusters.

The vacuum infrastructure of the micro-Newton thruster test facility consists of a vacuum tank (1500 L) which is equipped with the required pumps (forestage pump, two turbo pumps, one cryo pump) to enable a sufficient neutral gas background pressure during thruster testing. All pumps are decoupled via special isolator bellows from the vacuum vessel. The cryo pump used was customised to reduce the seismic noise impact of the pump to a minimum. The tank itself is also decoupled via optical isolator feeds from the floor. The noise shielding of the vacuum facility maintains the performance of the measurement instruments used and especially the performance of the developed thrust balance.

The key parameters of the facility are summarised in table 5.1. The pumping speed was chosen to maintain a neutral gas background pressure during operation of the tested EP thrusters in the 10^{-6} mbar regime which is sufficient to characterise the most important micro-Newton thruster technologies.

As main measurement instrument of the micro-Newton thruster facility, a highly precise and highly stable micro-Newton thrust balance has been developed, integrated, tested and used to characterise EP micro-Newton thrusters. The most important performance figures in comparison with the targeted values are presented in table 5.2. The table illustrates that all important development goals have been fulfilled. However, the balance is designed to be able to bear the maximum load and thus it can be assumed that it is possible to characterise heavier devices.

© Springer Fachmedien Wiesbaden GmbH 2018
F. G. Hey, *Micro Newton Thruster Development*,
https://doi.org/10.1007/978-3-658-21209-4_5

Table 5.1: Summary of the vacuum facility key parameters.

Facility Key Parameter	Specific Value
Vacuum chamber dimensions	1200 mm x 1200 mm x 880 mm
Pumping speed	11400 L/s
Background pressure without gas ballast	$2 \cdot 10^{-7}$ mbar
Background pressure with gas ballast (0.5 sccm)	$3 \cdot 10^{-6}$ mbar

As mentioned, the developed balance fulfils the targeted requirements, therefore the balance is able to characterise electric thrusters in the whole LISA measurement bandwidth.

According to the knowledge of the author of this thesis, this capability is not available anywhere else in the world. Especially the resolution of below 0.1 µN at very low frequencies has been not achieved by other thrust balances for micro-Newton thruster testing.

Table 5.2: Summary of the develop thrust balance performance.

Thrust Balance Parameter	Targeted Value	Achieved/Tested Value
Thrust range	0 µN to 2500 µN	0 µN to 4800 µN
Resolution	≤ 0.1 µN	≤ 0.1 µN
Measurement bandwidth	10 Hz to $2 \cdot 10^{-3}$ Hz	10 Hz to $1 \cdot 10^{-3}$ Hz
Thrust noise sensitivity	$0.1 \, \mu N / \sqrt{Hz}$	$\leq 0.1 \, \mu N / \sqrt{Hz}$
Maximum DUT weight	6 kg	4.4 kg

In addition to the thrust balance a plasma diagnostic setup had been developed, integrated and tested. The plasma diagnostic setup and thrust balance enable a detailed characterisation of the thruster under test. The diagnostic consists of a rotatable jib arm ($\pm 90\,°$), 15 Faraday cups and a Retarding Potential Analyser (RPA). The Faraday cups and the RPA are placed on the jib arm which allows an angular thruster plume characterisation. The key features and the performance data of the plasma diagnostic are summarised in table 5.3.

In the chapters 3.3 and 4 measurement results were presented that illustrate the unique balance performance for example the measured difference between direct and indirect total efficiency is in the order of some percent and is in agreement with the uncertainty analysis performed, which can be seen as an independent proof that the thrust balance and the plasma diagnostic delivers developed realistic data.

The thrust balance developed was also used to characterise the RITµX thruster which is currently the micro-Newton thruster with the highest TRL (see chapter 3, section 3.4). The results illustrate that the balance is able to characterise a real electric thruster because the commanded thrust levels were always in agreement with the measured thrust.

Table 5.3: Key features of the plasma diagnostic assembly

Plasma Diagnostics Key Parameter	Specific Value
Installed probes	15 Faraday cups & 1 RPA
Max. detectable current density	$< 2\,\mu A$
Retarding voltage range	$0\,kV - 6000\,kV$
Angular range	$\pm 90\,°$

The measured balance performance is illustrated in figure 5.1. The PSD shown, where the thrust (in µN) is normalised to the frequency (in $\sqrt{Hz}$) and logarithmically plotted, provides an overview of the noise level at specific frequencies. The dash-dot-dotted line represents the LISA requirement (see chapter 1), which is given by the required thrust level combined with the typical LISA allocation [7]. The dashed curve illustrates the Euclid and the NGGM requirement. The dotted curve represents the thrust noise performance of the thrust balance developed and shows that, the balance fulfils the LISA requirement in the complete measurement bandwidth. Additionally, the thrust balance fulfils the thrust noise requirement of Euclid and NGGM. Moreover, the overall measurement range of the balance covers the targeted throttle range of the thruster candidates for LISA, NGGM and Euclid. Id est, the thrust balance can be used to fully characterise possible micro-Newton thruster candidates for the targeted and other future scientific space missions. Thus, the identified gap of a missing highly stable thrust balance, which had been explained in the introduction of this thesis, can be seen as closed.

In figure 5.1 also the measured thrust noise performance of the tested RITµX system is presented as the dash-dotted curve. The direct thrust noise measurement of the RITµX system was performed for the first time. The thruster system almost fulfils the LISA thrust noise requirement, although the thruster system developed was not designed for an optimal thrust noise performance. Therefore, the measured RITµX system performance indicates strongly that the thruster system which is based on the RIT principle can fulfil the targeted thrust noise requirement of LISA.

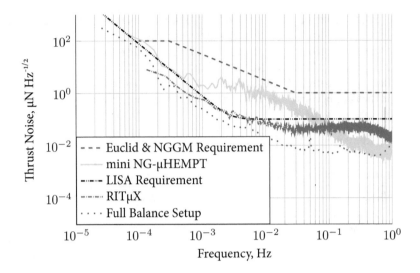

Figure 5.1: The plot presents the directly measurement thrust noise performances which
has been presented in the thesis. The dashed curve illustrates the requirements
of NGGM and Euclid. The dash-dot-dotted curve shows the requirement of
LISA. The solid curve is the measured thrust noise performance of the developed
mini NG-µHEMPT. The dash-dotted curve shows the measured thrust noise
performance of the RITµX that has been tested with the developed micro-
Newton thruster test facility. The performance of the thrust balance which is part
of the facility is presented as the dotted curve.

The development of a next generation µHEMPT was also presented in this thesis.
The thrusters were developed via a semi empirical development approach. This
approach was chosen to accelerate the development of the µHEMPT compared to
previous development efforts. Three different thrusters had been built. A prototype
was built which reused existing thruster parts of the EBB20 developed by A. Keller.
which was used to demonstrate that the assumptions on which the semi empirical
development approach relies were applicable.

Based on this success, two next generation µHEMPTs were developed, built
and characterised. A miniaturised version, that has a similar discharge chamber
diameter ($\leq$ 8 mm) as the already existing thrusters was used to achieved the targeted
development goals with reference to LISA.

To study the losses of the thruster downscaling, a model with a wider discharge
chamber diameter was developed, built and characterised in parallel. The so called
full scale model is 4 times larger than the miniaturised model, which a the maximum

discharge chamber diameter of 30 mm. All other parts of the thruster has been linear scaled as well.

The testable operation space of the full scale NG-μHEMPT was comparatively wide. Table 5.4 presents the ranges of the most important thruster parameters. The thruster could be operated at very low propellant mass flow levels and hence at very low thrust levels. The specific impulse and the total efficiency at these very low thrust levels was reduced.

The best total efficiency and the best specific impulse could be achieved with propellant mass flows between 2 - 3 sccm and an anode potential between 600 V and 1200 V. At this operation high specific impulses ($\geq$ 2000 s) and a total efficiency of up to 30 % could be measured. The maximum measured thrust was 4800 μN.

It is clearly evident, that the optimal operation points of the full scale thruster do not cover the targeted thrust regime.

Table 5.4: Summary of the thruster operation points of the full scale thruster. The thruster demonstrated a very stable operation over a wide operation range. The best total efficiency and the best specific impulse were achieved with propellant mass flows between 2 sccm and 3 sccm and an anode potential between 600 V and 1200 V.

Parameter	Value
Anode potential range	150 V - 4400 V
Anode current range	0.06 mA - 244 mA
Mass flow range	0.08 sccm - 4 sccm
Thrust range	6.3 μN - 4800 μN
I_{sp} range	120 s - 2100 s
PTTR range	20 W/mN - 61 W/mN
η_T range	1 % - 28.5 %

In parallel to the testing of the full scale thruster the characterisation of the miniaturised thruster developed has been presented and analysed in the previous chapters. An overview of the achieved operation range of the miniaturised NG-μHEMPT is given in table 5.5. The overall operation range of the thruster is not as wide as the operation range of the full scale model. The thruster is able to generate tens of micro-Newtons of thrust which is in agreement with the targeted thrust regime.

With 900 s the specific impulse of the thruster presents a significantly higher efficiency than the thruster that was developed until 2013. But the thruster does not fulfils the targeted specific impulse requirement of 1000 s.

In figure 5.1 the thrust noise performance of the miniaturised NG-μHEMPT developed is also presented as the yellow curve. It can be seen that the thruster does not fulfil the LISA thrust noise requirement over the whole measurement bandwidth. The currently non-compliant thrust noise performance could be caused by the preliminary anode current controller that was used to control the thruster during the measurement.

The developed miniaturised NG-μHEMPT is only with respect to the thrust level fully compliant with the targeted requirements. To be compliant with all other introduced requirements, a further continuation of the thruster development is necessary. However, the difference between the targeted values and achieved values is small for example 10 % in the case of the specific impulse. Therefore, a further development appears to be promising.

Table 5.5: Summary of the determined operation range of the miniaturised NG-μHEMPT. The thruster can be operated in the micro-Newton regime. At equal thrust levels the miniaturised thruster presents higher specific impulses than the full scale thruster.

Parameter	Value
Anode potential range	195 V - 2400 V
Anode current range	2.7 mA - 6 mA
Mass flow range	0.08 sccm - 0.9 sccm
Thrust range	29 μN - 86 μN
I_{sp} range	300 s - 900 s
PTTR range	29 W/mN - 82 W/mN
η_T range	2 % - 7 %

Based on the thruster development performed and the results achieved a further development of the miniaturised NG-μHEMPTs is planned. A further optimisation of the design of the miniaturised thruster based on the data obtained and a numerical simulation performed shall improve the overall thruster performance to be compliant with the presented requirements.

Additionally, it should be analysed if the existing thruster could be potentially used on a CubeSat, or comparable mission. Although the mini NG-μHEMPT is not as efficient as other technologies such as the RIT, it is 18 times more propellant efficient as a cold gas thruster system that has a comparable system complexity. Paired with the long lifetime that is featured by the HEMPT principle the thruster could also be able to replace a small cold gas system in a small science or Earth observation mission.

The laboratory model could also be used to develop a cold gas thruster that is able to operate in boosted mode, for mission scenarios which require very low thrust with a very low specific impulse and higher thrust levels paired with a higher specific impulse. The low specific impulse in very low thrust operation could be tolerated due the overall system mass (thruster system mass plus propellant mass).

With respect to LISA-like missions this would mean that the thruster that has to compensate the radiation pressure of the sun would operate in boosted mode, whereas the other thrusters which would be part of the AOCS could operate in cold gas mode. Such a system could allow an extremely power and mass efficient AOCS thruster operation.

In parallel to the further development of the miniaturised thruster, it is planned to use the full scale NG-μHEMPT as baseline thruster to develop and test a milli-Newton μHEMPT Elegant Bread Board (EBB) system model to demonstrate the capabilities of the full scale thruster as main thruster for small satellites. The promising performance of the full scale NG-μHEMPT with respect to small Hall-Effect thrusters and the low system complexity of a HEMPT system could be interesting for satellites with a mass of around 100 kg.

The total efficiency of the full scale NG-μHEMPT is a factor 1.5 to 2 higher than the total efficiency of comparable hall-effect thrusters, if they are operated in the same thrust regime [115]. An additional benefit of a milli-Newton μHEMPT system could be the low overall system mass compared to other EP technologies. To further simplify the thruster and its operation, it is imaginable to modify the magnetic field configuration of the thruster in order to enable a simplified ignition sequence.

The system EBB will include the thruster, a neutraliser, a power converter, a mass flow controller and additional pipe work. It is also planned to perform tests with iodine as an alternative propellant.

Currently, the next generation of the micro-Newton thrust balance is in development. The development is based on the experience which was made during the presented work. The main goal of the new development is an optimisation of the thermal design of the thrust balance paired with other minor changes.

A sketch of the new thrust balance generation, called Thrust Balance MkIV, is presented in figure 5.2. The new setup shall be able to maintain the full resolution and stability also in the case that several tens of watts are installed on the thrust balance. This is especially important to characterise more complex EP thrusters, like the RITμX, over the whole operation space of the tested thrusters.

As minor changes the mechanical design shall be modified to increase the resolution by a factor of 2. The structural modifications shall also lead to an improved insensitivity to seismic noise. Moreover, the Thrust Balance MkIV will be able to bear higher maximum thruster weights and also whole thruster system assemblies.

Based on the presented plasma diagnostic setup and the experience that was made, a new plasma diagnostic assembly will be developed and integrated. The new setup shall allow a measurement of the ion current and ion energy at every measurement

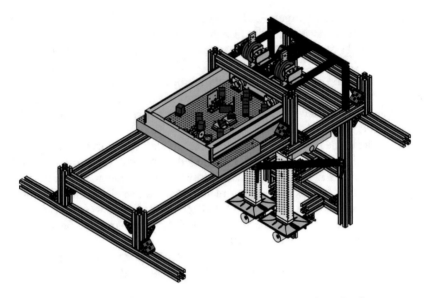

Figure 5.2: Illustration of the Thrust Balance MkIV that is currently in development. It is based on the experience which was made during the presented work. The baseline configuration has not changed, although the several performance and design improvements will be implemented. Especially the thermal design of the balance assembly will be optimised to increase the long term stability and the balance resolution. The targeted balance resolution is smaller than 0.05 µN over the whole LISA measurement bandwidth.

angle in two dimensions. The setup is shown in figure 5.3. It can be recognised that the new sensor will be arranged in a spherical configuration instead of the cylindrical configuration that is currently used. The new diagnostic will consists of 15 novel energy selective Faraday probes which can also be described as gridless retarding potential analyser. The probes are based on an idea that was formulated during the present activity and for which a patent is currently pending. The new setup will allow a spherical plume characterisation over $\pm 90°$ in one direction and $\pm 45°$ in the other angular direction. It is planned to maintain the proven stepper motor configuration and to change only the measurement instruments on the jib arm.

Together with the new probes, an improved electronics will also be implement based on digital data processing. The measured ion current shall be directly transferred into a digital signal at the sensors inside the vacuum chamber. Every single sensor will be connected to a digital bus to transfer the measured information to the workstation. The new electronics will also lead to an extended and selectable

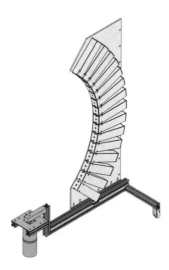

Figure 5.3: Illustration of the new plasma diagnostic assembly. It will consists of 15 novel energy selective Faraday probes which can also be described as grid less retarding potential analyser. The probes are based on an idea which is currently patent pending and that was invented during the presented work. The new setup will allow a spherically plume characterisation over $\pm 90°$ in one direction and $\pm 45°$ in the other angular direction.

measurement range (up to $200\,\mu A/mm^2$) without reducing the sensor resolution. The large measurement range will enable the characterisation of micro- and milli-Newton thruster with a single device.

In parallel to the integration of the novel energy selective Faraday probes in the micro-Newton thruster test facility some of the probes shall be tested in other EP test facilities to verify if the new instrument works as predicted.

With respect to the achieved overall performance of the micro-Newton thruster test facility, it is foreseen to continue the characterisation of thrusters developed internally and outside of Airbus. Especially in the scope of the LISA technology development roadmap the further characterisation of possible LISA AOCS thruster candidates will be continued.

Bibliography

[1]   Edgar Y. Choueiri. "A Critical History of Electric Propulsion: The First Fifty Years (1906-1956)". In: *AIAA-2004-3334* (2004).

[2]   Dan M. Goebel. *Fundamentals of Electric Propulsion*. John Wiley & Sons, 2008.

[3]   ECSS-E-ST-10C. *Space engineering - System engineering general requirements*. Tech. rep. European Cooperation for Space Standardization, 2009.

[4]   Marc D. Raymana, Thomas C. Fraschetti, Carol A. Raymond and Christopher T. Russell. "Coupling of system resource margins through the use of electric propulsion: Implications in preparing for the Dawn mission to Ceres and Vesta". In: *Acta Astronautica, Volume 60, Issue 10-11, p. 930-938* (2007).

[5]   Albert Einstein. "Zur allgemeinen Relativitätstheorie". In: *Sitzungsberichte der Königlich Preussischen Akademie der Wissenschaften; Berlin* (1915).

[6]   Albert Einstein. "Über Gravitationswellen". In: *Sitzungsberichte der Königlich Preussischen Akademie der Wissenschaften; Berlin* (1918).

[7]   European Space Agency. *Lisa, Unveiling a hidden Universe*. ESA/SRE, 2011.

[8]   Thilo Schuldt. "An Optical Readout for the LISA Gravitational Reference Sensor". PhD thesis. Humbolt Universität zu Berlin, 2010.

[9]   European Space Agency. *NGO Revealing a hidden Universe: opening a new chapter of discovery*. ESA/SRA, 2011.

[10] G. Heinzel, A. Ruediger and R. Schilling. *Spectrum and spectral density estimation by the Discrete Fourier transform (DFT), including a comprehensive list of window functions and some new at-top windows*. Tech. rep. Max-Planck-Institut fuer Gravitationsphysik (Albert-Einstein-Institut) Teilinstitut Hannover, 2002.

[11] Enrico Canuto, Andrés Molano-Jimenez, A. Bacchetta, M. Buonocore, S. Cesare, B. Girouart and L. Massotti. "The control challenges for the Next Generation Gravity Mission". In: *AIAA Guidance, Navigation, and Control Conference, Boston, MA* (2013).

© Springer Fachmedien Wiesbaden GmbH 2018
F. G. Hey, *Micro Newton Thruster Development*,
https://doi.org/10.1007/978-3-658-21209-4

[12]   European Space Agency. *Euclid, Mapping the geometry of the dark Universe, Assessment Study Report.* ESA/SRE, 2009.

[13]   C.V.M. Fridlund. "Darwin – The Infrared Space Interferometry Mission". In: *European Space Agency Bulletin* (2000).

[14]   Thales Alenia Space. *Euclid Micro-Propulsion System (MPS) REQUIREMENT SPECIFICATION EUCL-TAST-RS-2-051.* European Space Agency, 2014.

[15]   Elizabeth Buchen. "Small Satellites Market Observations". In: *29th Annual AIAA/USU Conference in Small Satellites* (2015).

[16]   WorldVu Satellites Ltd. *Satellites make it all possible.* Jan. 2016. URL: http: //oneweb.world/#solution (visited on 07/01/2016).

[17]   George Paul Sutton and Oscar Biblarz. *Rocket Propulsion Elements.* Willey, 2010.

[18]   Jürgen Müller. "Thruster Options for Microspacecraft: A Review and Evaluation of Existing Hardware and Emerging Technologies". In: *Jornal of Applied Physics* (2007).

[19]   J.Mitterauer. *Prospects of Liquid Metal Ion Thrusters for Electric Propulsion.* Tech. rep. IEPC-1991-105. Vienna University of Technology, 1991.

[20]   M. Tajmar, A. Genovese and W. Steiger. "Indium Field Emission Electric Propulsion Microthruster Experimental Characterization". In: *Journal of Propulsion and Power* 20.2 (Mar. 2004), pp. 211–218. ISSN: 0748-4658.

[21]   Daniel Bock, Alexander Kramer, Philip Bangert, Klaus Schilling and Martin Tajmar. "NanoFEEP on UWE platform - Formation Flying of CubeSats using Miniaturized Field Emission Electric Propulsion Thrusters". In: *IEPC-2015-121 / ISTS-2015-b-121* (2015).

[22]   Rainer Killinger, Howard Gray, Ralf Kukies, Michael Surauer, Giorgio Saccoccia, Angelo Tomasetto and Ray Dunster. "Artemis Orbit Raising In-Flight Experience with Ion Propulsion". In: *38th AIAA/ASME/ASEE Joint Propulsion Conference and Exhibit, Indianapolis, Indiana, July 2002* (2002).

[23]   G. Saccoccia. "ESA Spacecraft Propulsion Activities". In: *Proceedings of the 4th International Spacecraft Propulsion Conference (ESA SP-555).* (2004).

[24]   Norbert Koch, Günter Kornfeld and Gregory Coustou. *The HEMP Thruster - An Alternative to Conventional Ion Sources?* Retrieved: June 2015, Available at http://www.uni-leipzig.de/~iom/muehlleithen/2003/2003_ koch.pdf. Thales. Mar. 2003. URL: http://www.uni-leipzig.de/ ~iom/muehlleithen/2003/2003_koch.pdf.

[25]   Günter Kornfeld, N. Koch and G. Coustou. *First Test Results of the Hemp Thruster Concept*. Tech. rep. Thales Electron Devices GmbH, 2001.

[26]   Hans-Peter Harmann, Günter Kornfeld and Norbert Koch. "Low Complexity and Low Cost Electric Propulsion System for Telecom Satellites Based on HEMP-Thruster Assembly". In: *IEPC-2007-114* (2007).

[27]   Norbert Koch, Hans-Peter Harmann and Günter Kornfeld. "Development and Test Status of the Thales High Efficiency Multistage Plasma Thruster Familiy". In: (2005). IEPC-2005-297.

[28]   Norbert Koch, Stefan Weis, Martin Schirra, Alexey Lazurenko, Benjamin van Reijen, Jens Haderspeck, Angelo Genovese and Peter Holtmann. "Development, Qualification and Delivery Status of the HEMPT based Ion Propulsion System for SmallGEO". In: *IEPC-2011-148* (2011).

[29]   Stefan Weis, Alexey Lazurenko, Benjamin van Reijen, Jens Haderspeck, Angelo Genovese, Ralf Heidemann, Peter Holtmann, Klaus Ruf and Norbert Püttmann. "Overview, Qualification and Delivery Status of the HEMPT based Ion Propulsion System for SmallGEO". In: *IEPC-2015-345 / ISTS-2015-b-345* (2015).

[30]   Andreas Keller, Peter Köhler, Waldemar Gärtner, Benjamin Lotz, Davar Feili, Philipp Dold, Marcel Berger, Claus Braxmaier, Dennis Weise and Ulrich Johann. "Feasibility of a down-scaled HEMP-thruster". In: *IEPC-2011-138* (2011).

[31]   Andreas Keller, Peter E. Köhler, Franz Georg Hey, Marcel Berger, Claus Braxmaier, Davar Feili, Dennis Weise and Ulrich Johann. "Parametric Study of HEMP-Thruster Downscaling to μN Thrust Levels". In: *IEEE Transactions on Plasma Science* 43.1 (Jan. 2015), pp. 45–53. ISSN: 0093-3813.

[32]   Andreas Keller. "Feasibility of a down-scaled HEMP Thruster". PhD thesis. University of Gießen, 2013.

[33]   ECSS-E-ST35C. *Space engineering - Propulsion general requirements*. Tech. rep. European Cooperation for Space Standardization, 2008.

[34]   Peter E. Köhler and Bruno K. Meyer. "Beam Diagnostics for Mini Ion Engines". In: *IEPC-2013-297* (2013).

[35]   Denis Packan, Paul-Quentin Elias, Julien Jarrige, Félix Cannat, Clément Zaepffel, Julien Labaune and Trevor Lafleur. "Electric Propulsion Activities at ONERA". In: *IEPC-2015-29 / ISTS-2015-b-29* (2015).

[36]   Martin Tajmar and G. Fiedler. "Direct Thrust Measurements of an EMDrive and Evaluation of Possible Side-Effects". In: *Propulsion and Energy Forum*. American Institute of Aeronautics and Astronautics, 2015.

[37] Eduard Bosch Borràs, José González del Amo and Alexandra Bulit. "ESA Propulsion Laboratory at ESTEC". In: *IEPC-2015-60 / ISTS-2015-b-60* (2015).

[38] B. Seifert, A. Reissner, N. Buldrini, F. Plesescu and C. Scharlemann. "Development and Verification of a μN Thrust Balance for High Voltage Electric Propulsion Systems". In: *IEPC-2013-208* (2013).

[39] Kristof Holste, Waldemar Gärtnerrtner, Peter Köhler, Patrick Dietz, Jennifer Konrad, Stefan Schippers, Peter J. Klar, Alfred Müller and Peter R. Schreiner. "In Search of Alternative Propellants for Ion Thrusters". In: *IEPC-2015-320 / ISTS-2015-b-320* ().

[40] Hans-Peter Harmann. "Untersuchung und Modellierung der Ionenstrahlformung großflächiger Ionenquellen mit Hilfe einer beweglichen Faradaysondenzeile". PhD thesis. Justus-Liebig Universität Gießen (Fachrichtung Physik), 2003.

[41] Julien Jarrige, Denis Packan, Olivier Duchemin and Lahib Balika. "Assessment of the Azimuthal Homogeneity of the Neutral Gas in a Hall Effect Thruster using Electron Beam Fluorescence". In: *IEPC-2015-12 / ISTS-2015-b-12* (2015).

[42] Paul-Quentin Elias, Julien Jarrige, Edoardo Cucchetti, Denis Packan and Alexandra Bulit. "Full Ion Velocity Distribution Function measurement in an Electric Thruster, using LIF-based tomographic reconstruction". In: *IEPC-2015-235 / ISTS-2015-b-235* (2015).

[43] J. Jarrige, P. Thobois, C. Blanchard, P.-Q. Elias, D. Packan, L. Fallerini and G. Noci. "Thrust Measurements of the Gaia Mission Flight-Model Cold Gas Thrusters". In: *Journal of Propulsion and Power* (2014), pp. 1–10.

[44] Denis Packan, Jean Bonnet and Simone Rocca. "Thrust Measurements with the ONERA Micronewton Balance". In: *IEPC-2007-118* (2007).

[45] Bernhard Seifert, Alexander Reissner, Nembo Buldrini, Thomas Hörbe, Florin Plesescu, Alexandra Bulit and Eduard Bosch Borras. "Verification of the FOTEC μN Thrust Balance at the ESA Propulsion Lab". In: *IEPC-2015-258ISTS-2015-b-258* (2015).

[46] J. Perez Luna. "Development Status of the ESA Micro-Newton Thrust Balance". In: *IEPC-2011-011* (2011).

[47] Eduard Bosch Borràs, José González del Amo and Ben Hughes. "ISO17025 Accreditation of the ESA Micro-Newton Thrust Balance". In: *IEPC-2015-259 / ISTS-2015-b-259* (2015).

[48] Franz Georg Hey. "Development, Integration and Test of a Micro Newton Thrust Balance". MA thesis. Techische Universität Dresden, 2012.

[49]   Harald Koegel. "Interferometric characterization and modeling of path-length errors resulting from beamwalk across mirror surfaces in LISA". In: *Appl Opt* (2013).

[50]   Martin Gohlke. "A High Sensitivity Heterodyne Interferometer as a Possible Optical Readout for the LISA Gravitational Reference Sensor and its Application to Technology Verification". In: *7th International LISA Symposium* (2009).

[51]   Filippo Ales, Oliver Mandel, Peter Gath, Ulrich Johann and Claus Braxmaier. "A phasemeter concept for space applications that integrates an autonomous signal acquisition stage based on the discrete wavelet transform". In: *Review of Scientific Instruments* 86.8, 084502 (2015), pp. -.

[52]   Gerald Hechenblaikner, Ulrich Johann, Marc-Peter Hess, Marcel Berger, Stefan Winkler and Raphael Naire. "Challenges in spacecraft design and measurement strategies for future fundamental science missions and instruments". In: *39th COSPAR Scientific Assembly. Held 14-22 July 2012, in Mysore, India. Abstract H0.3-6-12, p.738* (2012).

[53]   Thilo Schuldt, Martin Gohlke, Dennis Weise, Ulrich Johann, Claus Braxmaier, Ruven Spannagel and Harald Koegel. "Picometer-Interferometry as a Key Enabling Technology for Future Space Missions". In: *39th COSPAR Scientific Assembly. Held 14-22 July 2012, in Mysore, India. Abstract H0.3-13-12, p.1732* (2012).

[54]   D. Gerardi, G. Allen, J. W. Conklin, K-X. Sun, D. DeBra, S. Buchman, P. Gath, W. Fichter, R. L. Byer and U. Johann. "Invited Article: Advanced drag-free concepts for future space-based interferometers: acceleration noise performance". In: *Review of Scientific Instruments* 85.1, 011301 (2014), pp. -.

[55]   John W. Dankanich, Michael W. Swiatek and John T. Yim. "A Step Towards Electric Propulsion Testing Standards: Pressure Measurements and Effective Pumping Speeds". In: *AIAA 2012-3737* (2012).

[56]   Richard Blott, Steven Gabriel and David Robinson. *Draft Handbook for Electric Propulsion (EP) Verification by Test (Issue 0)*. ESA/ESTEC, 2012.

[57]   Hans-Peter Harmann, Günter Kornfeld and Norbert Koch. "Physics and Evolution of HEMP-Thrusters". In: *IEPC-2007-108* (2007).

[58]   A. von Keudell. *Einführung in die Plasmaphysik*. Ruhr-Universität Bochum, 2014.

[59]   M. Tajmar. "Overview of Indium LMIS for the NASA-MMS Mission and its Suitability for an In-FEEP Thruster on LISA". In: *IEPC-2011-009* (2011).

[60] John K. Ziemer. "Sub-Micronewton Thrust Measurements of Indium Field Emission Thrusters". In: *IEPC-2003-247* (2003).

[61] Hans Leiter. "Entwicklung, Bau und Test eines RIT15 „Breadboard Engineering Models"". PhD thesis. University of Gießen, 2000.

[62] Hans Leiter and Rainer Killinger. "Development of the Radio Frequeny Ion Thruster RIT XT – A Status Report". In: *IEPC-01-104* (2001).

[63] Hans-Peter Harmann, Günter Kornfeld and Norbert Koch. "Status of the THALES High Efficiency Multi Stage Plasma Thruster Development for HEMP-T 3050 and HEMP-T 30250". In: *IEPC-2007-110* (2007).

[64] Günter Kornfeld, Gregory Coustou and Norbert Koch. US 7,247,992 B2. 2007.

[65] Günter Kornfeld, Gregory Coustou and Norbert Koch. DE 10300776. 2003.

[66] Hans-Peter Harmann, Norbert Koch and Günter Kornfeld. "Plasmabeschleunigeranordnung". DE 10 2006 059 264. 2006.

[67] Günter Kornfeld, Gregory Coustou and Norbert Koch. DE 103 18 925. 2003.

[68] Günter Kornfeld, Hans-Peter Harmann and Norbert Koch. DE 10 2007 043 955. 2007.

[69] Günter Kornfeld, Gregory Coustou, Werner Schwertfeger and Lenz Roland. DE 101 30 464. 2001.

[70] Jens Haderspeck, Stefan Weis, Benjamin van Reijen, Angelo Genovese, Alexey Lazurenko, Ralf Heidemann, Peter Holtmann, Klaus Ruf and Norbert Püttmann. "HEMP Thruster Assembly Performance with increased Gas Tubing Lengths of Flow Control Unit". In: *IEPC-2015-346 / ISTS-2015-b-346* (2015).

[71] Günter Kornfeld. "A Fully Kinetic and Self-Consistent Simulation of a μN-HEMP-Thruster Using Random Cell Scattering (RCS) for Solving the Änomalous Electron Transport" Problem". In: *IEPC-2015-406 / ISTS-2015-b-406* (2015).

[72] Tim Brandt, Claus Braxmaier, Frank Jansen, Thomas Trottenberg, Holger Kersten, Franz Georg Hey, Ulrich Johann and Rodion Groll. "Simulation for an improvement of a down-scaled HEMP thruster". In: *IEPC-2015-374 / ISTS-2015-b-374* (2015).

[73] James E. Polk, Anthony Pancottiy, Thomas Haagz, Scott Kingx, Mitchell Walker, Joseph Blakelyk and John Ziemer. "Recommended Practices in Thrust Measurements". In: *IEPC-2013-440* (2013).

[74] Herbert Balke. *Einführung in die Technische Mechanik, Kinetik.* Springer, 2005.

[75] Manuel Gamero-Castañom, Vlad Hruby and Manuel Martínez-Sánchez. "A Torsional Balance that Resolves Sub-micro-Newton Forces". In: *IEPC-01-235* (2001).

[76] Enrico Canuto and Andrea Rolino. "Nanobalance: an automated interferometric balance for micro-thrust measurement". In: *ISA Trans* 43.2 (2004), pp. 169–87. ISSN: 0019-0578.

[77] K. Marhold and M. Tajmar. "Direct Thrust Measurement of In-FEEP Clusters". In: *IEPC-2005-235* (2005).

[78] Stefano Cesare, Fabio Musso, Filippo D'Angelo, Giuseppe Castorina, Marco Bisi, Paolo Cordiale, Enrico Canuto, Davide Nicolini, Eliseo Balaguer and Pierre-Etienne Frigot. "Nanobalance: the European balance for micro propulsion". In: *IEPC-2009-182* (2009).

[79] Nathaniel, P. Selden, Andrew and D. Ketsdever. "Comparison of force balance calibration techniques for nano-Newton range". In: *Review of Scientific Instruments* (2003).

[80] M. B. Hopkins and W. G. Graham. "Langmuir probe technique for plasma parameter measurement in a medium density discharge". In: *Review of Scientific Instruments* 57.9 (1986), pp. 2210–2217.

[81] Benjamin van Reijen, Norbert Koch, Alexey Lazurenko, Stefan Weis, Martin Schirra, Angelo Genovese, Jens Haderspeck and Eberhard Gill. "High Precision Beam Diagnostics for Ion Thrusters". In: *IEPC-2011-132* (2011).

[82] C. Bundesmann, C. Eichhorn, F. Scholze, H. Neumann, H. J. Leiter, D. Pagano and F. Scortecci. "Electric Propulsion Thruster Diagnostic Activities at IOM". In: *IEPC-2015-392ISTS-2015-b-392* (2015).

[83] Thomas Trottenberg, Alexander Spethmann and Holger Kersten. "An Interferometric Force Probe for Thruster Plume Diagnostics". In: *IEPC-2015-419ISTS-2015-b-419* (2015).

[84] *S-2000 StabilizerTM Vibration Isolators and Optical Table Installation Manual.* Newport. 2010.

[85] Facility Effects in Stationary Plasma Thruster Testing. "T. Randolph and V. Kim, and H. Kaufman and K. Kozubsky and V. Zhurin and M. Day". In: *IEPC 1993-093, 23rd Int. Electric Propulsion Conference* (1993).

[86] A. Sengupta, J. A. Anderson, C. Garner, J. R. Brophy, K. L. deGroh, B. A. Banks and T. A. Karniotis Thomas. "Deep Space 1 Flight Spare Ion Thruster 30,000 Hour Life Test". In: *Journal of Propulsion and Power Vol. 25* (2009).

[87] Hans Leiter, Dagmar Bock, Benjamin Lotz, Davar Feili and Clive Edwards. "Qualification of the miniaturized Ion Thruster RIT μX - Perspectives, Program, Results and Outlook". In: *IEPC-2011-316* (2011).

[88] *Vacuum Technology Book.* Pfeiffer Vacuum GmbH, 2009.

[89] Gerd Jakob and Jean-Louis Lizon. "Low-vibration high-cooling power 2-stage cryocoolers for ground-based astronomical instrumentation". In: *SPIE 7733-139 Astronomical Telescopes and Instrumentation* (2010).

[90] J.L. Lizon, G. Jakob, B. de Marneffe and A. Preumont. "Different ways of reducing vibrations induced by cryogenic instruments". In: *Proceedings of the SPIE, Volume 7739, id. 77394B* (2010).

[91] Waldemar Gärtner. "Design, Konstruktion und Test eines präzisen Gegenfeldanalysators zur energetischen Strahlvermessung eines μN RIT Triebwerkes". MA thesis. Justus-Liebig Universität Gießen Arbeitsgruppe: Ionentriebwerke, 2010.

[92] Andreas Keller, Franz Georg Hey, Peter Koehler, Marcel Berger, Claus Braxmaier, Davar Feili, Dennis Weise and Ulrich Johann. "Parametric Study of HEMP-Thruster, Downscaling to μN Thrust Levels". In: *IEPC-2013-269* (2013).

[93] K. Dannenmayer, S. Mazouffre, M. Merino and E. Ahedo. "Hall Effect Thruster Plasma Plume Characterization with Probe Measurement and Self-Similar Fluid Models". In: *48th AIAA/ASME/SAE/ASEE Joint Propulsion Conference & Exhibit* (2012).

[94] Christian Böhm and Jérome Perrin. "Retarding-field analyzer for measurements of ion energy distributions and secondary electron emission coefficients in low-pressure radio frequency discharges". In: *Review of Scientific Instruments* (1993).

[95] Maximilian Schramm. "Development of Thruster Plume Diagnostic Tools for Micro High Efficiency Multistage Plasma Thruster Characterization". MA thesis. Institut für Raumfahrtsysteme der Universität Stuttgart, 2014.

[96] Herbert Balke. *Einführung in die Technische Mechanik, Festigkeitslehre.* Springer, 2008.

[97] C.-M. Wu, S.-T. Lin and J. Fu. "Heterodyne interferometer with two spatial-separated polarization beams for nanometrology". In: *Optical and Quantum Electronics* 34.12 (2002), pp. 1267–1276. ISSN: 1572-817X.

[98] E. Morrison, B. J. Meers, D. I. Robertson and H. Ward. "Experimental demonstration of an automatic alignment system for optical interferometers". In: *Appl. Opt., 33(22):5037–5040* (1994).

[99]   Andreas Neumann, Julian Sinske and Hans-Peter Harmann. "The 250mN Thrust Balance for the DLR Goettingen EP Test Facility". In: *IEPC-2013-211* (2013).

[100]   William A. Johnson and Larry K. Warne. "Electrophysics of Micromechanical Comb Actuators". In: *Journal of Microelectromechanical Systems*, 4.1 (Mar. 1995), pp. 49–59. ISSN: 1057-7157.

[101]   A. Jamison, A. Ketsdever and E. P. Muntz. "Gas dynamic calibration of a nano-newton thrust stand". In: *Review of Scientific Instruments* (2002).

[102]   E. Hering, R. Martin and M. Stohrer. *Taschenbuch der Mathematik und Physik*. Springer, 2004.

[103]   Davina Di Cara, A Bulit, J. Gonzalez del Amo, J.A. Romera, H. Leiter, D. Lauer, C. Altmann, R. Kukies, A. Polli, L. Ceruti, A. Antimiani, D. Feili and B. Lotz. "Experimental Validation of RIT Micro-Propulsion Subsystem Performance at EPL". In: *IEPC-2013-269* (2013).

[104]   D. Di Cara, L. Massoth, F.Musso S. Cesare, G. CasG. Castorina, D. Feili and D. Lotz. "Performance verification of the µN RIT-2.5 thruster on the Nanobalance facility". In: *IEPC-2011-013* (2011).

[105]   C. Altmann, H. Leiter and R. Kukies. "The RIT-µX Miniaturized Ion Engine System way to TRL 5". In: *IEPC-2015-274ISTS-2015-b-274* (2015).

[106]   Tim Brandt, Thomas Trottenberg, Rodion Groll, Frank Jansen, Franz Georg Hey, Ulrich Johann, Holger Kersten and Claus Braxmaier. "Particle-in-Cell simulation of the plasma properties and ion acceleration of a down-scaled HEMP-Thruster". In: *4th International Spacecraft Propulsion Conference*, (2014).

[107]   Tim Brandt, Thomas Trottenberg, Rodion Groll, Frank Jansen, Franz Georg Hey, Ulrich Johann, Holger Kersten and Claus Braxmaier. "Simulations on the influence of the spatial distribution of source electrons on the plasma in a cusped-field thruster". In: *The European Physical Journal* (2015).

[108]   HKCM Engineering e.K. *Sm2CO17, YXG32 Specification*. Tech. rep. HKCM Engineering e.K., 2016.

[109]   HKCM Engineering e.K. *NdFeB, 42UH Specification*. Tech. rep. HKCM Engineering e.K., 2016.

[110]   Vacuumschmelze GmbH & Co. KG. *Selten - Erd - Dauermagnete VACOdym / VACOmax*. Tech. rep. Vacuumschmelze GmbH & Co. KG, 2016.

[111]   K. Matyash, O. Kalentev, R. Schneider, F. Taccogna, N. Koch and M. Schirra. "Kinetic Simulation of the staionary HEMP Thruster including the-field plume region". In: *IEPC-2009-110* (2009).

[112]   N. Koch, M. Schirra, S. Weis, A. Lazurenko, B. van Reijen, J. Haderspeck, A. Genovese, P. Holtmann, R. Schneider, K. Matyash and O. Kalentyev. "The HEMPT Concept - A Survey on Theoretical Considerations and Experimental Evidences". In: *IEPC-2011-236* (2011).

[113]   K. Matyash, R. Schneider, A. Mutzke, O. Kalentev, F. Taccogna, N. Koch and Martin Schirra. "Kinetic Simulations of SPT and HEMP Thrusters Including the Near-Field Plume Region". In: *Plasma Science, IEEE Transactions on* 38.9 (2010), pp. 2274–2280.

[114]   FuG Elektronik GmbH. *Hochspannungsnetzgeräte Serie HCP von 3,5 kV bis 300 kV / 14 W bis 15000 W*. Tech. rep. FuG Elektronik GmbH, 2016.

[115]   Tommaso Misuri, Cosimo Ducci, Riccardo Albertoni, Mariano Andrenucci and Daniela Pedrini. "Sitael Low Power Hall Effect Thrusters for Small Satellites". In: *IEPC-2015-102ISTS-2015-b-102* (2015).

[116]   C. Ducci, S. Oslyak, D. Dignani, R. Albertoni and M. Andrenucci. "HT100D performance evaluation and endurance test results". In: *IEPC-2013-140* (2013).

Appendix

Summary: Numerical Electro Static Comb Displacement Survey

A finite element method simulation was performed in order to determine the calibration error of the Electro Static Comb (ESC) assembly. The simulation had been performed with Ansys Maxwell in three dimensions, therefore the impact of displacements along X, Y and Z as well as rotations around X, Y and Z was assessed. To reduce the amount of data processed, a set of parameter had been defined. In figure 1 the axis-definition used is shown. The information presented summarises only the cases with the highest impact on the relative calibration error which is typically a combination of displacements, rotations or displacements plus rotations along the axes. However, the simulation had been performed for various parameter sets including displacements or rotations around or along a single axis. It had been also verified that a displacement along the X-axis has a small or negligible impact on the calibration error, as described in the literature.

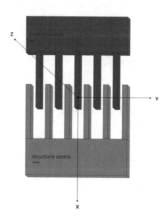

Figure 1: Illustration of a single Electro Static Comb (ESC) including the coordinate system used

© Springer Fachmedien Wiesbaden GmbH 2018
F. G. Hey, *Micro Newton Thruster Development*,
https://doi.org/10.1007/978-3-658-21209-4

In the following the combined displacement and rotation in all degrees of freedom of the ESC is presented. It has been found that this case is the worst case and leads therefore to the highest relative calibration errors. For this case the following maximum and minimum numbers had been used:

• Minimal displacment: 0.05 mm with minimal rotation angle 0.5 deg

• Maximal displacment: 0.45 mm with maximal rotation angle 4.5 deg

Figure 2 summarises the result for the combination of displacement and rotation, it can be seen that a displacement larger than 0.15 mm coupled with a rotation of 1.5 deg leads to a relative error of 3.6 %. Due the Airbus ESC alignment procedure

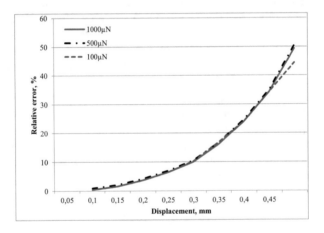

Figure 2: Illustration of the relative error plotted against the displacement coupled with the rotation for different applied forces.

defined and used, the assembly's maximal displacement is typically smaller than 0.125 mm and smaller than 1.25 deg in all degrees of freedom. Thus, the maximum relative error is smaller than 3 %.

To illustrate the exclusive influence of the rotation, in the following an overview is presented, where the ESC had been tilted around the X-, Y- and Z-axis. Figure 3 summaries the results. It has been found that a rotation around all axis simultaneously is the worst case for the calibration error. In summary the minimum and maximum rotation values used are:

• Minimal rotation angle: 0.5 deg

• Maximal rotation angle: 4.5 deg

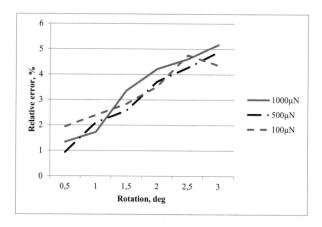

Figure 3: Illustration of the relative error plotted against the rotation for different applied forces.

The following provides the worst case calibration error for a displacement along the X-, Y-, and Z-axis. Figure 4 shows the results, where the calibration error is plotted versus the displacement. A simultaneous displacement along all axis leads to the highest calibration error, although the impact of a displacement along the X-axis the small. In summary the minimum and maximum displacement values used are:

- Minimal displacement: 0.025 mm

- Maximal displacement: 0.3 mm

The displacement has a stronger influence on the relative error of the ESC than the rotation error. The maximum error is smaller than the maximum error of the combined error case (rotational error and displacement).

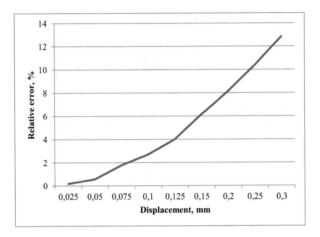

Figure 4: Illustration of the relative error plotted against the displacement.

BILLIONS OF CONNECTIONS HAVE ONE THING IN COMMON. AIRBUS.

WE MAKE IT FLY

Our customers include all the major satellite operators – delivering video, data and internet to homes, offices or people on the move – as well as solutions for military and government organisations too. We have the right technology they need. We offer the flexibility to adapt to changing markets. And our electric satellites allow more cost-effective missions and greater capacity to answer growing business needs.

Together. We make it fly.

airbus.com **AIRBUS**

Printed in the United States
By Bookmasters